Social and Cultural Aspects of VCR Use

COMMUNICATION

A series of volumes edited by:
Dolf Zillmann and **Jennings Bryant**

Zillmann/Bryant • Selective Exposure to Communication

Beville • Audience Ratings: Radio, Television, Cable, Revised Edition

Bryant/Zillmann • Perspectives on Media Effects

Goldstein • Reporting Science: The Case of Aggression

Ellis/Donohue • Contemporary Issues in Language and Discourse Processes

Winett • Information and Behavior: Systems of Influence

Huesmann/Eron • Television and the Aggressive Child: A Cross-National Comparison

Gunter • Poor Reception: Misunderstanding and Forgetting Broadcast News

Olasky • Corporate Public Relations: A New Historical Perspective

Donohew/Sypher/Higgins • Communication, Social Cognition, and Affect

Van Dijk • News Analysis: Case Studies of International and National News in the Press

Van Dijk • News as Discourse

Wober • The Use and Abuse of Television: A Social Psychological Analysis of the Changing Screen

Kraus • Televised Presidential Debates and Public Policy

Masel Walters/Wilkins/Walters • Bad Tidings: Communication and Catastrophe

Salvaggio/Bryant • Media Use in the Information Age: Emerging Patterns of Adoption and Consumer Use

Salvaggio • The Information Society: Economic, Social, and Structural Issues

Olasky • The Press and Abortion, 1838–1988

Botan/Hazleton • Public Relations Theory

Zillmann/Bryant • Pornography: Research Advances and Policy Considerations

Becker/Schoenbach • Audience Responses to Media Diversification: Coping With Plenty

Caudill • Darwinism in the Press: The Evolution of an Idea

Richards • Deceptive Advertising: Behavioral Study of a Legal Concept

Flagg • Formative Evaluation for Educational Technologies

Haslett • Communication: Strategic Action in Context

Rodda/Grove • Language, Cognition and Deafness

Narula/Pearce • Cultures, Politics, and Research Programs: An International Assessment of Practical Problems in Field Research

Kubey/Csikszentmihalyi • Television and the Quality of Life: How Viewing Shapes Everyday Experience

Kraus • Mass Communication and Political Information Processing

Dobrow • Social and Cultural Aspects of VCR Use

Barton • Ties That Blind in Canadian/American Relations: Politics of News Discourse

Bryant • Television and the American Family

Cahn • Intimates in Conflict: A Communication Perspective

Biocca • Television and Political Advertising, Volume 1: Psychological Processes

Welch • The Contemporary Reception of Classical Rhetoric: Appropriations of Ancient Discourse

Hanson/Narula • New Communication Technologies in Developing Countries

Social and Cultural Aspects of VCR Use

Edited by
Julia R. Dobrow
Boston University

LAWRENCE ERLBAUM ASSOCIATES, PUBLISHERS
1990 Hillsdale, New Jersey Hove and London

Lawrence Erlbaum Associates, Inc., Publishers
365 Broadway
Hillsdale, New Jersey 07642

Library of Congress Cataloging-in-Publication Data

Social and cultural aspects of VCR use / edited by Julia R. Dobrow.
p. cm.—(Communication)
Includes bibliographical references.
ISBN 0-8058-0499-4
1. Video tape recorders and recording—Social aspects—United States. I. Dobrow, Julia R. II. Series: Communication (Hillsdale, N.J.)
HE8700.6.S65 1990
303.48′33—dc20 89-28205
CIP

Printed in the United States of America
10 9 8 7 6 5 4 3 2 1

Contents

Contributors

Stanley J. Baran • *Department of Theatre Arts, San Jose State University*
Julia R. Dobrow • *College of Communication, Boston University*
Cheryl Harris • *Department of Communication, University of Massachusetts/Amherst*
Katharine E. Heintz • *Department of Communication, University of Illinois Champaign/Urbana*
Amy B. Jordan • *Department of Communication, Widener University*
Bruce C. Klopfenstein • *Department of Radio-Television-Film, Bowling Green State University*
Megumi Komiya • *Telecommunications Consultant, London, England*
Carolyn A. Lin • *Department of Radio-TV, Southern Illinois University at Carbondale*
Barry Litman • *Department of Telecommunication, Michigan State University*
Kimberly K. Massey • *Department of Communication, University of Utah*
Michael Morgan • *Department of Communication, University of Massachusetts/Amherst*
Eugene Secunda • *Department of Marketing, Baruch College, CCNY*
James Shanahan • *Department of Communication, University of Massachusetts/Amherst*
Joseph D. Straubhaar • *Department of Telecommunication, Michigan State University*
Lawrence J. Vale • *Department of Architecture, Massachusetts Institute of Technology*

Introduction

Julia R. Dobrow
Boston University

Although commercially available in the United States for more than a decade, videocassette recorder (VCR) sales continue to rise. Not surprisingly, therefore, interest in VCR use and its possible effects has not waned, on the part of both the academic community and media industries. The number of scholarly papers and articles about video increase each year. So do the number of articles on video in various trade publications. What's all the fuss about?

In addition to undertaking some of the pioneering studies about video use in this country, Levy (1987, 1989) has written cogently about the place of VCR research in the field of mass communication. Although concluding that video has not created a "revolution" in the ways in which people use media or in providing great diversity in what people view, Levy nonetheless has suggested that part of the interest in video comes from it providing a "fresh site" to test out some older theories, to see how video is or isn't like other technologies, and to explore the effects of video on different individuals or groups of people.

Meanwhile, the VCR has generated great interest among those in the various media industries. Television and film analysts want to know how much of a threat video poses to their respective shares of the entertainment audience. Advertising executives are concerned about the extent to which VCRs enable people to avoid ads and how this can be prevented, and those in the ratings industries have tried to cope with how to measure viewing in an age of video. Communications lawyers have struggled with the difficult issues of copyright and intellectual property created by the taping and replaying capabilities of VCRs. The rapid growth of video in the United States has also spawned a whole series of subsidiary industries: video stores, video magazines, video accessories, and videocameras. Newspaper and magazines include charts of best-selling videocassettes, offer tips for VCR purchase, maintenance, or advice on what to tape from the local TV fare. In a manner similar to what Marc (1984) termed "self- reflexive television," television programs themselves have bitten the video bullet and make frequent (if somewhat ironic) reference to VCRs in various storylines. There is no doubt about it: The VCR is a compelling focal point for people interested in media.

Whereas the "first generation" of studies about VCRs sought to quantify (how many VCRs were out there) and categorize (who tended to buy them, for what functions were they most often used and with what frequency), the "second generation" of research on video focused more on patterns of use and the social and psychological gratifications derived from it. As Levy (1989) noted in his historiography of VCR work, researchers began to explore some of the questions about the extent to which the VCR fits into previously identified patterns of other media use, what light VCR use might shed on other theories, and indeed, the question of whether of not VCR use should be considered a form of mass communication.

It is possible that this is the pattern of research that has followed the introduction of every new communication technology: Numbers and demographic pattern studies precede those studies about patterns of use, uses and gratifications, and other effects studies. Only after a technology has been established and penetrated a substantial percentage of the households in a given country, are researchers able to take a step back and try to assess the technology's place within a social and/or cultural context. To be sure, some of the earlier studies on VCR use touched on these issues. Studies conducted in the late 1980s started to focus on them.

This volume contains some of the "third generation" of writing about video. Although several of the chapters continue to address the very important questions raised in the previous two generations of VCR research, the authors here have sought to explore how the VCR fits into a larger social and cultural framework. Now that video is firmly entrenched in American society (as well as many others) and within many cultural subgroups, they collectively ask, can we re-examine the technology and see what it means? Has the VCR, in any fundamental way, had impact on some of our cultural institutions? What is its place among other media industries? What is its place in our homes and families? What is its place in our social lives, and what is its role in defining our place within our culture?

There are many common threads running through the chapters in this book. Almost all of the chapters in some way focus on the issue of how the VCR has given the viewer a degree of control: control over exposure (or re-exposure) to specific programming, control over who can see what, and control over time. Many of the chapters explore the ways in which VCRs are like books—the extent to which they are used selectively, actively, and individualistically; the extent to which people use stories on videotape the way they use written ones; and even the extent to which videotapes are treated like (and stored alongside) books.

Interestingly, many of the authors discuss ways in which video is similar to audio technologies: Ratings systems that take VCRs into account should be conceived of as more like radio audience ratings; selective exposure with video is more like selecting audio content; re-exposure via video may be analogous to re-listening to favorite tapes, records, or CDs. Indeed, it might turn out that VCR use owes at least as much to audio technologies as it does to ones with pictures.

Although some of the chapters in this book deal with the ways in which video is or may be altering our home environments, having some impact on our socialization or resocialization or our self-image, others focus on video as a symbol of change in our social environments and the ways in which we might present ourselves to others.

In a sense, this book is about *relationships*. The chapters in Part I discuss the relationship of the VCR to other media industries. Secunda tells the story of the birth and growth of the VCR in the United States. Amidst highly competitive and protective television and film industries, the VCR stood out as the technology that would allow for viewer control and, Secunda argues, threatened the very core of those other industries upon which the VCR depended. Komiya and Litman follow up along this line, discussing the economics of the videocassette industry. They discuss how the VCR forced changes in film production, distribution, and duplication, and give an in-depth analysis of the videocassette market structure. Klopfenstein's chapter focuses on the relationship of the VCR to the various audience ratings industries. He shows that very different ratings can be achieved—and are being achieved—by using different methodologies, and he suggests some reasons for the discrepancies.

The chapters in Part II examine a wide range of issues, but all attempt to see the relationship of the VCR to various theoretical frameworks. Lin examines the concept of audience activity as it applies in VCR use, and finds that people with VCRs are active "users" rather than passive "viewers." She questions whether the choices the VCR affords its users will create "idiosyncratic home video or media subcultures." Massey and Baran agree that the VCR certainly allows for choices. They consider the degree to which VCRs liberate people from the programming schedules, and ask whether the VCR has had any impact on ways in which people spend their leisure time.

In addition to freeing the audience from television programming schedules, the VCR was also heralded as the technology that would allow people virtually unlimited choices in content, giving each VCR owner control of his or her programming destiny. But would this freedom mean people would expose themselves to individualized messages, taking them away from the stories we all know from TV, disseminator of much common culture? Is the promise of diversity fulfilled? Morgan, Shanahan, and Harris examine this issue, and apply the theory and methods of cultivation analysis to the case of the VCR.

Common exposure to certain cultural symbols, such as the ones shown on television, has had many functions within societies. One function has been the development of taste: another has been the reinforcement of class stratification. Straubhaar questions whether video will have any impact on these issues. Using examples collected from fieldwork in several Latin American countries, as well as other examples from all over the world, he focuses on video as an indicator of social class and taste.

The chapters in Part III look at the relationship of the VCR to the behavior and use patterns of individuals. Each chapter addresses questions of what patterns of video use by individuals can tell us about other social and cultural issues. Heintz looks at VCR libraries as opportunities for parents to control what their children view. Jordan utilizes ethnographic analysis to explore the way video use fits into family life. Both of these studies address questions of how the VCR is used, symbolically, as a point of control and, practically, as a source of socialization.

My own chapter focuses on the question of repetitive behavior. Clearly the VCR makes repeat viewing easy, but why do people do it—what personal, social, and iconographic functions might it play? Finally, Vale looks at the camcorder, the technology which enables individuals to construct and preserve their worlds on video. He discusses what kinds of things are filmed and why, and what it might mean to people to see themselves on television.

Taken together, these chapters make only a small dent in the possible spectrum of video research. There is clearly much more work to be done. As others have noted throughout more than a decade of video research, it will be important to follow individuals and groups with VCRs longitudinally, for only then will we be able to assess trends and accurately discern patterns of video use and what they mean. It will also be important to continue to chart the impact of VCRs on other media industries and to explore the impact of video on other social and cultural institutions, including individuals, family, peer, and reference groups. We must continue to investigate what distinctions in patterns of video use and meaning can be made based on class, ethnicity, gender, age, or other social factors. We need to assess not only the literal roles and functions that VCRs and videocassettes play within a given cultural context, but also what symbolic functions they might have. Cross-cultural and interdisciplinary research can only enrich the field.

A volume such as this is the product of collective effort. I am grateful to the contributors for the hard work and thought which made this book possible. The many graduate students who assisted me in data collection, and especially Hillary Brooks, Isabelle Delbos, Doris Gottlieb, Pi-Tzu Kuo, and Margaret Wallace, deserve a hearty "thank you" for their efforts. Thanks are due also to Todd Martini, Robin Marks Weisberg, and Hollis Heimbouch at Lawrence Erlbaum Associates. Jennings Bryant gave me tremendous support and feedback from the inception of this book through its final stages. I am grateful to George Gerbner for starting me out on a topic I now feel I've finished. My colleague and friend Nick Mills deserves special thanks for his support. And finally, as always, emotional support and love from my parents and my husband was enormously influential and greatly appreciated.

Julia R. Dobrow

REFERENCES

Levy, M. R. (1989). Why VCRs aren't pop-up toasters: Issues in home video research. In M. R. Levy (Ed.), *The VCR age*. Newbury Park, CA: Sage.

Levy, M. R. (1987). Some problems of VCR research. *American Behavioral Scientist, 30*(5), 461–470.

Marc, D. (1984). *Demographic vistas*. Philadelphia, PA: University of Pennsylvania Press.

I

The Relationship of VCRs to Other Media Industries: Competition, Cooperation, and Confusion

1

VCRs and Viewer Control Over Programming: An Historical Perspective

Eugene Secunda
Baruch College

Groups of American television executives crowded into a small Chicago hotel room on April 21, 1956 to witness the public introduction of the Ampex videotape recorder. As they watched the demonstration, there was no indication that they had any idea of the enormous impact this new media technology would ultimately have throughout the world. To most of the professional broadcasters in attendance, it was just a brilliant solution to a myriad of production and scheduling problems. At the time, it was impossible for these industry leaders to envision the transition of this technology from broadcast studio to living room some 20 years later when Sony Corporation would introduce the Betamax videocassette recorder to America.

In 1956, television was emerging lustily from its infancy. With TV sets in almost 72% of all U.S. households ("Trends in television," 1987), the medium had already established itself as a dominant cultural influence in this country. However, since its birth as a commercial medium following World War II, the industry struggled to deal with the constraints of having to do live telecasts (Barnouw, 1975).

Before the Ampex system was introduced, the television networks had to broadcast virtually all of their programming live. The exception was feature-length movies, previously produced for theatrical exhibition.

Almost every important television show originated from New York City in the mid-1950s. That meant TV audiences on the U.S. west coast were at a disadvantage because of the time difference. They received the shows (produced in the evening on the east coast) 3 hours earlier—in mid-afternoon, when most of viewers were still at work. Scheduling problems were further compounded because daylight savings time was not yet in effect uniformly throughout the United States ("Birth of a new," 1956).

The professional adoption of the Ampex videotape recorder transformed the entire television industry. It meant that programs (with the exception of sports events) could be prerecorded at the convenience of the producers, transmitted to affiliated stations via coaxial cable, then telecast locally at the same "clock time" throughout the country. Two decades later, this "time-shifting" concept was embraced by millions of Americans who then had the capability to videotape TV programs in their own homes and play them back at their own convenience. In essence, it permitted every VCR owner to become a television programmer (Brown, 1979).

The professional introduction of the videotape recorder (a technology that permits the recording of television signals on magnetic tape for later replay) augered the end of television watching as a video version of what Ong (1982, p. 74) described as an "oral culture." Before the advent of the videotape recorder, both the presenter and the listener had to be in attendance at the time a program was telecast. Ong said that television and other related electronic technologies ushered in the age of "secondary orality," bearing significant similarities to the era before literacy was generalized, "in its participatory mystique, its fostering of a communal sense, its concentration on the present moment. . . ." (p. 136).

The spontaneity and tension usually associated with live performances were sacrificed as scenes were endlessly re-taped in an effort to achieve a sense of perfection. It is possible that this new taped TV era laid the groundwork for public acceptance, some 20 years later, of the home videotaping of television programming as a surrogate for watching live action.

The videotaping technology that allowed millions to gain control of the medium actually originated in Denmark in 1898. Valdemer Poulsen, a physicist, proposed a theory he called "telegraphon" that involved the use of magnetic recording to capture sound on wire. Poulsen's hypothesis was extended and patented in England by Boris Rtcheouloff in 1927. He specifically suggested that this idea could be adapted to record television signals on magnetized substances (Abramson, 1955).

The German industrial firm, I.G. Farben, is credited with being the first to actually produce technology based on Poulsen's theory. The company began manufacturing audio recorders, labeled *magnetophons,* in 1931 (Schubin, 1986). These recorders captured and replayed sound with significantly less noise interference than other existing recording devices of the time. They were used by Nazi Germany during World War II. Impressed with their superior recording and playback capability, U.S.

Army electronics technician, John T. Mullin, brought several of the units home from Europe with him in 1945.

Two years later, famed singer and network radio star Bing Crosby heard about these machines and invited Mullin to tape his shows magnetically for later broadcast on the ABC radio network. These broadcasts, in 1947, marked the first time that magnetic tape had been used professionally in the United States to time-shift programming. The success of this experiment inspired Crosby to commission the Ampex Electric Corporation to develop improved tape recorders. Thus, a new electronics industry was launched in this country, positioning America as a leader in magnetic tape recorder technology for the next several decades (Mullin, 1976).

Ampex, previously an aircraft engine producer, became the dominant force in this new field. The entire radio industry embraced the new technology, liberating itself from the stringent limitations of live broadcasting. At the same time, Ampex engineers were beginning to transfer the knowledge they accumulated during the development of the audio recorder to the burgeoning area of video. The California company demonstrated its first working magnetic tape video recorder and playback mechanism in 1951 (Winston, 1986).

By that time, television was in nearly 25% of all U.S. homes ("Trends in television," 1987). RCA, both a manufacturer of consumer electronics products and a major participant in the rapidly expanding network TV industry, also began the pursuit of videotape recorder research. RCA produced a working prototype of its system in 1953, but it was proven impractical and was never sufficiently re-developed to reach the marketplace. Meanwhile, Ampex successfully demonstrated its videotape recorder in 1956 and it was quickly embraced as the TV industry's standard, thereby neutralizing competitors' efforts (Adams, 1956).

The Ampex system, based on the use of large open reels containing 2-inch wide magnetic tape was the standard videotape recorder design for the next 14 years. The concept utilized a transverse scanning system that recorded electronic signals in a zig-zag pattern across the surface of magnetic tape (Lardner, 1987). The American company was soon established worldwide as the primary supplier of videotape recording equipment for professional markets. But it essentially ignored the technology's potential for home use.

Whereas Ampex concentrated on producing and selling professional equipment, other electronics companies in the United States, Europe, and Japan began to pursue videotape recording for the consumer market. In the early 1950s, The Sony Corporation, led by chairman Akio Morita, established a goal of producing and marketing home video recorders. Sony and other Japanese manufacturers were directly supported in this endeavor when the Japanese government organized an industry-wide approach to further research and development in the field. By 1959, Sony, Matsushita Electric Corporation (Sony's chief rival), and other Japanese companies had developed working prototypes of home videotape recorders using a different method than that established by Ampex. Called the helical scanning system, this method rejected Ampex's 2-inch wide standard and allowed more information to be recorded on the tape. Continuous improvements have been made in videotape recording since that time. However, the helical scanning system, developed in Japan during the 1950s, remains the conceptual basis for virtually all videocassette recorders currently being produced (Lardner, 1987).

Sony began marketing its helical scanning VTRs to Americans in the early 1960s. One of its first customers was American Airlines; they used the Japanese equipment to screen Hollywood movies for passengers flying between the U.S. east and west coasts ("Airline TV," 1964). The field was soon crowded with scores of competitors who sensed the home video market's potential and were determined to exploit it.

Wesgrove Electrics and Telcan from Britain, Philips from the Netherlands and Loewe-Opta from West Germany all created prototype home video recorders and announced plans for the introduction of their products to the American consumer market. In the United States, the ITT Research Center, Par Ltd., and Fairchild Camera and Instrument produced early entries in the still untested product category. In 1963, Ampex belatedly realized the possibilities in the consumer market, but its first videotape recorder for home use was priced at a substantial $30,000 (Lardner, 1987). Two years later, Ampex returned to the marketplace with a more realistically priced unit for the home ($1,095 retail), however, its bulk and limited playing time did not garner significant buyer response ("Video recorder becomes," 1965).

As increasing numbers of Japanese and European consumer electronics companies began exploring the prospects of the untapped American home video market, many major U.S. companies also began taking notice. Instead of responding with aggressive research and development programs (as the Japanese were doing) to create patented home video technology, companies like General Electric and Concord Electronics Corp. began positioning themselves primarily as marketers of video recorders manufactured for them by their Japanese competitors ("Two more VTRs," 1966). This trend, established in the 1960s, would ultimately cause the United States to lose its position as the dominant force in consumer electronics technical development. Today, in the largest and richest consumer market in the world, U.S. consumer electronics companies function almost exclusively as promoters and distributors of Japanese and Korean products bearing such familiar brand names as RCA, Zenith, and General Electric ("Home VTRs under," 1977). However, knowledgeable shoppers for videocassette recorders and other consumer electronics products are increasingly interested in knowing which non-U.S. companies actually manufactured the products bearing the famous American names.

In 1969 a final American effort was made to retain control of videocassette recorder manufacture and marketing. Cartridge TV, Inc., an amalgam of Playtape and Avco Industries, was formed to offer the American public a 1/4-inch videotape cassette system called "Cartrivision." The product was manufactured for Cartridge TV, Inc. by several different American television set manufacturers, and sold throughout the country by Sears Roebuck and Montgomery Ward department store chains. It was withdrawn from the market in 1973 after the company experienced a variety of production, marketing, and financial difficulties ("Cartridge TV halts," 1973). Ironically, the ability to record television shows was a primary promotional point when the unit was offered to consumers in the early 1970s. This appeal failed to produce significant sales (DeLuca, 1980). A few years later, Sony's advertisements for the Betamax VCR used the same sales approach, and that product was an immediate success in the U.S. market.

As American consumer electronics manufacturers surrendered the industry initiative to the Japanese in the late 1960s and early 1970s, three leading Japanese manufacturers (Sony, Matsushita, and JVC) started collaborating on videocassette recorder research (Nayak & Ketteringham, 1986). Representatives of the three companies signed an agreement to share all future technological innovations in the VCR field. Such an agreement would have been impossible among U.S. competitors because of the rigid legal restraints against monopolies established by our government. The cooperation of these Japanese companies helped them accelerate the progress of each one's VCR research and development efforts, although in later years they again became intense competitors.

While many Japanese firms were working on VCR research, Sony took an initial lead within the consumer electronics industry when it perfected a 3/4-inch videotape cassette system it called "U-Matic." Early in its marketing program Sony made some tentative gestures toward positioning the U-Matic as a VCR for home use but its bulk, high price ($2,500), and limited playing capacity restricted its potential for that purpose. However, it did achieve great success in professional applications and was ultimately established as the dominant audio–visual tool in the business, broadcast and educational sectors around the world (Morita, 1986). It was also the first successfully marketed videotape recorder using a cassette container to house magnetic tape instead of using the previously accepted method of holding the tape on exposed reels.

VCR VERSUS VIDEODISC

During the early 1970s, European and U.S. consumer electronics companies continued to seek ways to maintain primacy in an industry they had dominated since the advent of commercial radio in America. Almost by coincidence, these occidental giants chose the videodisc as the primary video technology with which to challenge Japan's domination of the global consumer video marketplace.

In 1972, Dutch-owned Philips Electronics and MCA, a U.S. entertainment company, announced the integration of their two similar video systems. They used laser beam technology to retrieve audio and video signals from a disc (comparable in size to a 33 1/3 phonograph record). They created a device that produced a video picture superior to any the VCRs could generate and was capable of storing enormous amounts of information that could be recalled on demand. The result of this collaboration was marketed in America under the brand name, "DiscoVision" ("DiscoVision closely," 1972). Subsequently, Philips and MCA were confronted (by RCA, the U.S. consumer electronics giant) with the introduction of a competitive videodisc system called "SelectaVision" (Videodisc battle," 1975). Both systems provided consumers with playback-only technology. This limitation was an increasingly apparent marketing liability because the Japanese had become successful selling American videotape cassette machines with both recording and playback capability.

These large European and American corporations, both marketing videodisc systems, battled each other throughout the late 1970s and early 1980s. But the American public never embraced either in a significant way; the appeal of the VCR's recording capability was just too strong. And apparently, the somewhat inferior picture quality provided by even the best VCR was not bad enough to dissuade American consumers from choosing the magnetic tape format over the videodisc. Although owners of VCRs clearly valued their ability to tape shows from television, a 1988 study by J. Walter Thompson Co. ("VCR: A new star in the media galaxy," 1988) revealed that twice as many hours are spent viewing prerecorded cassettes than recordings of TV programs. In 1987, industry publications reported that a re-recordable videodisc for the consumer market was under development. If such a product were to be introduced and mass marketed for home use at prices comparable to the VCR, it would be difficult to prophesize whether or not the technology could overcome the VCR's enormous marketing lead and challenge it as the preferred American home video format.

During the late 1960s and early 1970s there seemed to be general agreement within the consumer electronics industry that home video (in some form) would inevitably be part of the American media environment. The major issue was whose patented concept would become standard in the industry. Twenty non-compatible home video systems had been announced, and prototypes introduced by 1970 ("Videoplayers," 1970). Each competing company complained of lack of cohesiveness in the process of establishing an industry standard, and each proclaimed that its particular system was the one the consumer electronics business should ultimately accept. Ironically, the Federal anti-trust laws (originally established early in the 20th century to encourage free market competition) acted to stifle the pooling of American corporate research and development resources at the very same time the Japanese government (unhindered by similar legislation) was encouraging just such efforts. The result of these imbalanced policies was evident when Japanese-developed technologies were successfully established as a U.S. and global standard for VCR manufacture.

VHS VERSUS BETAMAX

The final contest for domination of the VCR consumer market began in the early 1970s while Japanese companies were still trying to persuade Americans that the cumbersome 3/4-inch U-Matic tape recorder was viable as a home video appliance. In 1971, when the management of JVC (a subsidiary of the giant Matsushita Electrical Industry Company, Inc.) concluded that Sony's 3/4-inch U-Matic format was not going to be a successful consumer product, they began research on a new 1/2-inch VCR concept, unrelated to previous designs. JVC (an acronym for Victor Company of Japan, Ltd.) was started by the U.S. owned RCA Corporation at the end of World War II, but was sold to Matsushita in 1953 (Nayak & Ketteringham, 1986). The JVC development team's understanding that emphasis must be concentrated on the unit's ability to record and playback for several hours was crucial in explaining JVC's later success as the creator of the dominant 1/2-inch VCR technology for the U.S. consumer market. Sony's U-Matic allowed for only 1 hour of use per cassette.

JVC rejected the U-Matic as a basis for further development (although they had total access to it under existing patent agreements), but Sony chose to use its design structure to accommodate the technical requirements of the 1/2-inch VCR. It was this design, created by Sony engineer Nobutoshi Kihara, that actually led to the production and initial success of the "Betamax" VCR in the U.S. market (Lardner, 1987). Sony's management, however, continued to believe that the majority of potential customers would be content with one hour of recording and playback capability. This view proved devastatingly wrong, as JVC and other competitors in the consumer electronics field later established.

Sony completed work on the Betamax in 1974 and began preparations for its launch into the U.S. market. At the same time, the company attempted to persuade its primary competition, JVC and its parent company, Matsushita, to drop their plans to develop a rival VHS system and instead, to embrace the Betamax system. They refused, and although they were slightly behind Sony's development timetable, they started to plan for a major U.S. introduction of their own VCR (Nayak & Ketteringham, 1986).

In November 1975, a little more than a year after Sony and its Japanese rivals terminated negotiations for a unified 1/2-inch VCR design concept, Sony introduced the Betamax VCR in America. During the previous decade, scores of other magnetic video tape recorder manufacturers had tried to convince potential American purchasers that each had produced the one VCR that would establish a U.S. standard. However, it was not until Sony launched the Betamax that American consumers finally made the commitment that created a whole new category in the consumer electronics industry. It was the beginning of the U.S. home video revolution (Lyons, 1976).

Although Sony originally announced that its initial entry in the U.S. market would be a Betamax VCR/19-inch color TV combination, priced at $2,295 ("Sony aims," 1975), the first VCR model actually offered Americans was a $1,300 Betamax tape deck that could be plugged into an existing television set. Advertising heralded the Betamax as a "video time-shift machine" for recording and playback of broadcast television programs ("Beta intro," 1976). This time-shift concept was the idea of Sony's chairman Akio Morita. It was the basis on which the VCR was to be sold to consumers. "With the VCR, television is like a magazine," Morita (1986) said. "You can control your own schedule" (p. 208).

In 1974, the inability of Sony, JVC, and Matsushita to agree on a single 1/2-inch VCR standard, led to the polarization of the Japanese and American consumer electronics industries. Both factions sought manufacturing and marketing partners. They each hoped the weight of their respective numbers could preempt a dominant market position early in the contest, thus forcing their rivals to capitulate. While pressing ahead with its own marketing efforts in the United States, Sony was simultaneously pursuing licensing agreements for the Betamax technology with other Japanese and American consumer electronics manufacturers and marketers. Early Betamax supporters were Japan's Toshiba and Sanyo and the Zenith Corporation in the United States. Soon after, champions of VHS announced that four U.S. companies, RCA, Magnavox, Sylvania, and Sears Roebuck, agreed to market VCRs in America using that format ("Magnavox, Sylvania," 1977).

Realizing they had misjudged market demand, Sony's management finally announced that they (along with their Betamax licensees) would begin U.S. distribution of VCRs with a 2-hour recording and playback capability. But by that time, the VHS forces were promoting machines with a 4-hour capability to the American market ("Japan moves," 1977). By 1977, a year and a half after Sony had introduced and popularized the concept of home video taping in America, 15 different VCR brands (utilizing both the Betamax and VHS formats) were being offered in the United States ("Home VTRs under," 1977). Although the VCR had begun to capture the attention and imagination of the public and the media, only 209,000 U.S. households had VCRs—significantly less than 1% of the total U.S. TV homes ("Trends in VCR usage," 1987).

VCRs AND PRERECORDED VIDEO

VCRs clearly gained initial consumer acceptance because they allowed viewers to record and playback television programs off the air. However, with the launch of the prerecorded video business in 1977, a second powerful reason to own a VCR materialized. Andre Blay, a Detroit businessman, started the new industry when he acquired from Twentieth Century Fox Film Corporation the rights to sell a number of its movies on videotape. He formed a video distribution company called Video Club of America and promoted mail order sales through national magazines. The concept caught on immediately, further spurring sales of VCRs to Americans eager to see recently released Hollywood feature films at home without waiting for them to appear on television (Lardner, 1987).

Only a few months after Andre Blay's innovative video sales business became an obvious success, video retail stores began proliferating throughout the United States. Videos of Hollywood films were initially offered to the viewers on a "for sale" only basis. But retailers soon began renting them to customers who balked at paying the high purchase price ($70–$80), but were willing to pay $10 or less for one night's showing ("Home video's pioneers," 1987). The extraordinary success of the video rental business frustrated the managements of the Hollywood studios. Although they had originally been happy to license video distributors to sell prerecorded copies of their films, as the volume of prerecorded video rentals grew, they began looking for ways to maintain tighter control over these rentals. Hollywood's concern about its inability to curb the home video industry led to a series of legal actions against the industry that continue through the late 1980s.

As the film companies and the home video industry battled, VCR sales in the United States accelerated. Sony and its Betamax allies were still leading the race for supremacy in the 1/2-inch VCR market, but were gradually beginning to lose ground to manufacturers and marketers of VHS-format machines. Recording capacity was the key to broad consumer acceptance; makers of VHS-type VCRs consistently offered more than their Betamax-rivals, eventually providing the capability to record and playback 6 hours of programming. By 1979, Americans were buying almost three times as many VHS-format VCRs than those utilizing the Betamax system ("VHS 71," 1979).

THE BETAMAX CASE

At the same time Sony was fighting its Japanese manufacturing counterparts for sales in the United States, it was forced to defend itself in a legal battle with several Hollywood studios. This action had a threatening effect on the company's ability to market its VCRs in America under any circumstances. Although Sony was the specific target of this litigation, the studios were actually challenging the legal right of any American to own and use a home videotape recorder regardless of the manufacturer's brand.

What has come to be called "The Betamax Case" began a year after Sony introduced its 1/2-inch VCR to the United States. MCA/Universal Studios and Walt Disney Productions filed a lawsuit in a California Federal District Court in 1976, charging the Japanese company with copyright infringement. They alleged that Sony (through its Betamax VCR advertising) was encouraging Americans to use Betamax machines illegally to videotape television shows produced and owned by the two studios ("Copyright suit," 1976). Legally or illegally, Sony's Betamax advertising was clearly promoting the concept that viewers owning Betamax VCRs were capable of choosing the time they watched their favorite television shows. The headline of a typical Betamax VCR advertisement read, "Make Your Own TV Schedule. Sony's Betamax can automatically videotape your favorite shows for you to play back anytime you want" ("Make your own," 1976).

By promoting the time-shifting methodology, then providing the means with which to accomplish it, Sony not only began the process that ultimately led a majority of America's television households to own (and increasingly use) a media technology that could wrest control of program flow from the three major television networks. It also weakened the near stranglehold Hollywood studios had maintained (over distribution of their product) since the earliest years of the film industry.

Sony was specifically selected as the target of the studios' legal wrath for two reasons:

1. Sony was the first company to successfully promote a 1/2-inch VCR technology in the United States. Until Sony's Betamax entered the market, no other VCR company had succeeded in capturing the imagination (and dollars) of the U.S. consumer. At the time the two studios brought their suit, Sony's Betamax was the only 1/2-inch VCR generally being sold in America.

2. The aggressive manner in which Sony (through its advertising) had encouraged Americans to videotape shows from television and view them at their convenience was regarded by MCA/Universal and Walt Disney Studios as an incitement to infringe on their copyrighted material.

The case see-sawed back and forth through the federal district court system for the next 8 years. Sony won the first round (Lindsay, 1979), but the District Court's decision was reversed in U.S. Appeals Court in 1981. The court found that "videotape recording of copyrighted television programs, even if it is done at home only for private use, is an infringement of rights of those who own programs" (Feder, 1981, p. 1).

This legal reversal elicited cheers from the Hollywood community and stunned the consumer electronics industry. Every company that manufactured or marketed VCRs for the U.S. consumer market, every advertising agency that promoted them and all the retailers who sold them, were now technically vulnerable to law suits. It also opened the possibility that every American using a VCR in his or her home to record television shows could theoretically be targeted for a copyright infringement suit and could be sued for damages.

In 1981, the battle between the pro- and anti-VCR forces moved to the political arena of Washington, DC. Both sides established powerful lobbying organizations to persuade Congress to legislate in favor of their respective positions. Sony, no longer alone in the fight, was joined by every other major consumer electronics manufacturer; they formed a "Right to Tape" coalition. The Motion Picture Association of America, already active as the chief exponent of the film industry's cause in Washington, accepted the responsibility of representing not only the MCA/Universal and Disney point of view on this volatile matter, but that of all the other studios as well (Lardner, 1987).

MCA/Universal and the other studios had originally focused their attention on aborting Americans' ability to videotape their television shows off the air. By 1981, however, the Hollywood community was far more concerned about finding ways to suppress the expanding U.S. videotape rental business, or at least to harness it so film producers could gain greater financial benefit. They petitioned sympathetic legislators to introduce bills providing the film industry with greater power to regulate the distribution of feature films on videotape. However, their lobbying efforts were neutralized by the VCR coalition. The "Right to Tape" group persuaded its allies in the House of Representatives and the Senate to introduce bills urging the government to "stop intruding into the homes of millions of Americans." No legislation directly affecting these issues ever came to a vote by either house of Congress ("Senate hearings," 1981).

In the midst of this furor, the U.S. Supreme Court declared it would review the 1981 ruling of the California Appeals Court. In January 1984, 2 years after taking the case under advisement and 8 years after it had been instituted, the Supreme Court reversed the Appeals Court decision. They found in favor of consumers' right to videotape television shows for their own use and decided that companies making or selling VCRs were not violating federal copyright law (Greenhouse, 1984). It was a devastating setback for the Hollywood studios; they were effectively preempted from controlling consumers' ability to copy studio works. It also established a legal precedent that discouraged the motion picture industry from pursuing any other schemes that would permit its member companies from levying royalties on the sale of blank videotape or VCRs in the future (Scherick, 1987).

1984 TO THE PRESENT

By the end of 1984 (12 months after the Supreme Court handed down its decision), VCR sales to consumers were proliferating throughout America. According to industry statistics, more than 7.5 million units were sold to retail dealers during that year, 86% more than in 1983 ("State of the industry," 1985). Budget-priced VCRs, largely imported from Korea, were available for as little as $200, one-sixth of what they had cost when they were introduced to the United States 9 years before. Videocassette recorder penetration of U.S. television households had reached 10.5% ("January street," 1985).

Hollywood's view of this phenomenon was unquestionably conflicted. In 1980, the film studios' total revenue derived from sales of prerecorded videotapes was a paltry $20 million (less than a single, successful Hollywood movie could gross in a few weeks). This sum represented only 1% of the studios' total income. In 1983, the film industry earned $625 million from videotaped movie sales, representing 14% of its total revenue (Scherick, 1987, p. 64). Producers of major box office hits, (referred to as "A" titles in the home video industry) were earning huge amounts from sales to prerecorded videocassette distributors. For example, George Lucas' "The Empire Strikes Back" reaped $12 million when it was sold for distribution to the home video market ("Competition looks on," 1984).

While overall VCR sales continued to grow, Sony's Betamax VCR format was obviously losing ground to VHS-type machines in the United States. Only 19% of all VCRs sold to American dealers in 1983 were Betamaxes ("VCR share," 1984). Realizing it was losing the 1/2-inch VCR format battle to VHS in the United States, Sony began a fighting retreat from the market category. In 1985, the company announced it would introduce a new and more compact VCR format—8mm—heralded in its promotional material as "the complete home video system of the future." At the time, Sony's spokesmen inferred that this new format would be the successor to Betamax, in essence suggesting that Betamax (and all other 1/2-inch VCR formats) would become obsolete ("Sony commits," 1985).

IMPLICATIONS

The videocassette recorder was a media technology born in the 1950s, responding to the needs of a burgeoning U.S. television industry. Twenty years later it was on its way to being adopted as a new appliance in the American home. The VCR's initial appeal to the consumer market was its ability to liberate the time when viewers watched television. It subsequently became equally popular because it allowed people to watch prerecorded tapes of first-run Hollywood films uninterrupted by commercials (as on broadcast television) and without paying monthly subscription fees for cable's pay-TV services.

Until the VCR became generally accepted in U.S. TV households, Americans' television viewing patterns were largely controlled by the dictates of the telecasters' program schedules. If viewers wanted to see particular programs, they had to be in front of a TV set at the time the show was broadcast. Further, viewing habits were largely dictated by the programming policies of the three broadcast television networks (especially during prime-time hours). The networks seduced audiences to stay tuned to their programs by strategically positioning them in a sequence most likely to encourage continuous viewing of that network throughout the evening. This technique, called *program flow,* was successfully utilized until viewers started feeling liberated by the increasing number of choices offered by independent stations, by cable TV and by the VCR.

However, it was not until the home video industry fully blossomed that Americans began to change their entertainment consumption habits profoundly. The video rental store became as ubiquitous as the supermarket in America's neighborhoods. VCR owners quickly adopted the idea of picking up a recently released Hollywood movie on the way home from work. And, instead of going to local movie theaters, young people began to gather in friends' homes on weekends to watch both new and classic motion pictures distributed on tape by motion picture studios.

The VCR, in tandem with other new electronic media technologies like cable TV, brought about dramatic changes in U.S. media consumption habits. The VCR in particular provided American viewers with the means to liberate themselves from the rigid time constraints imposed by the program schedules of traditional broadcast, as well as satellite delivered, television programming. Despite Sony's introductory exhortations to "add hours to your day," by automatically videotaping television shows, the VCR has not yet completely fulfilled its marketers' assumption that it would become a time-shift machine whose primary purpose is to free the viewer from actually being present when a video event is telecast. It is possible that the next, more technologically oriented generation, will fulfill that expectation as the complexity of handling the VCR is reduced and the new generation's mechanical sophistication grows. There is ample evidence already that the era of watching TV by the clock is passing.

REFERENCES

Abramson, A. (1955, February). A short history of television recording. *The Journal of the Society of Motion Picture and Television Engineers,* pp. 250–251.

Airline TV off the ground. (1964, July 6). *Television Digest,* p. 4.

Adams, V. (1956, April 15). TV is put on tape by new recorder. *The New York Times,* pp. 1, 76.

Barnouw, E. (1975). *The tube of plenty.* New York: Oxford University Press.

Beta intro in NY, $1300. (1976, February 16). *Television Digest,* p. 9.

Birth of a new medium—Home video recording. (1956, April 21). *Television Digest,* p. 9.

Brown, L. (1979). *Keeping your eye on television.* New York: The Pilgrim Press.

Cartridge TV halts production. (1973, July 9). *Television Digest,* p. 7.

Competition looks on. (1984, December 24). *Time,* p. 53.
Copyright suit challenge, Beta sales. (1976, November 15). *Television Digest,* p. 7.
DeLuca, S. M. (1980). *Television's transformation: The next 25 years.* San Diego: A. S. Barnes & Co., Inc.
DiscoVision closely resembles Philips VLP. (1972, December 18). *Television Digest,* p. 7.
Feder, B. J. (1981, October 20). Private videotaping of copyrighted TV rule infringement. *The New York Times,* sec. 1, p. 1.
Greenhouse, L. (1984, January 18). Television taping at home is upheld by the Supreme Court. *The New York Times,* sec. 1, p. 1.
Home video's pioneers: In their own words. (1987, August 17–21). *Twice,* pp. 32–33.
Home VTRs under 15 brand names. (1977, June 27). *Television Digest,* p. 8.
January street prices. (1985, February 4). *Television Digest,* p. 11.
Japan moves into technology lead. (1977, March 14). *Television Digest,* p. 9.
Lardner, J. (1987). *Fast forward: Hollywood, the Japanese and the VCR wars.* New York: W. W. Norton.
Lindsay, R. (1979, October 3). Home video recorders ruled lawful by judge. *The New York Times,* sec. 3, p. 11.
Lyons, N. (1976). *The Sony vision.* New York: Crown.
Magnavox, Sylvania choose 4-hour VHS. (1977, May 30). *Television Digest,* p. 7.
Make your own TV schedule. (1976). [Advertisement for Sony Betamax.] Sony Corporation of America.
Morita, A., with Reingold, E. M., Shimomura, M. (1986). *Made in Japan.* New York: E. P. Dutton.
Mullin, J. T. (1976, April). Creating the craft of tape recording. *High Fidelity Magazine,* pp. 62–67.
Nayak, R. P., & Ketteringham, J. M. (1986). *Breakthroughs.* New York: Rawson Associates.
Ong, W. J. (1982). *Orality and literacy: The technologizing of the word.* London: Methuen & Co.
Scherick, E. J. (1987, November). Crossed circuits. *American Film,* pp. 63–64.
Schubin, M. (1986, July). One man's history of television. *Videography.* pp. 49–50.
Senate hearings this week on taping bills. (1981, November 30). *Television Digest,* p. 7.
Sony aims for luxury market. (1975, May 5). *Television Digest,* p. 9.
Sony commits to 8mm as successor to 1/2" VCR. (1985, April 22). *Television Digest,* p. 11.
State of the industry. (1985, January 14. *Television Digest,* p. 10.
Trends in television. (1987, June). *Television Bureau of Advertising* (TvB), p. 3.
Trends in VCR usage. (1987, July). *Television Bureau of Advertising* (TvB), p. 3.
Two more VTRs. (1966, February 14). *Television Digest,* p. 12.
VCR: A new star in the media galaxy. (1988 November). *J. Walter Thompson Co. Media Resources & Research,* p. 2.
VCR share survey shows major realignment. (1984, March 12). *Television Digest,* p. 12.
VHS 71, Beta 20. (1979, July 9). *Television Digest,* p. 11.

Video recorder becomes consumer product. (1965, July 5). *Television Digest*, p. 7.
Videodisc battle-lines firmly drawn. (1975, March 4). *Television Digest*, pp. 7–8.
Videoplayers—20 systems in search of a market. (1970, July 20). *Television Digest*, p. 7.
Winston, B. (1986). *Misunderstanding media*. Cambridge, MA: Harvard University Press.

2

The Economics of the Prerecorded Videocassette Industry

Megumi Komiya
Barry Litman
Michigan State University

In the early 1980s, an executive at JVC (the major Japanese manufacturer of videocassette recorders that devised the VHS tape format) predicted that the penetration of VCRs would reach 70% in many countries. His rosy forecast was dismissed by many. Since then, sales of VCRs in the United States have grown nearly as fast as those of color television sets, from 2 million annual units in 1981 to over 12 million since 1986 (*Video Week,* 1983–1989). Today, VCRs can be found in over 65% of U.S. TV households,[1] and a large percentage of consumers have already purchased their second generation VCR (National Association of Broadcasters, 1989; "Second VCR," 1987). At a time when the bloom has faded from consumer demand for cable and home satellites, consumers are increasingly attracted to a program delivery system, the VCR, whose features and availability of software put them in total control of their media consumption, no longer subject them to a "temporal tyranny" dictated by network and local station programming executives.

More importantly, over-the-air and cable television are no longer the sole source of programming entertainment; viewers can now rent prerecorded videocassettes. Even though Hollywood motion pictures still dominate the pre-recorded videocassette offering, other subcategories of products such as music videos, "how-to's," classics, children videos, and adult-oriented material are forcefully establishing their market presence.

In view of the fact that the structure of the prerecorded videocassette

[1]The rate of VCR penetration is even higher in many foreign countries, especially those without cable television.

(PRVC) industry emulates that of the motion-picture industry, it is suggested that this industry may experience the same increasing degree of market concentration and vertical integration historically endemic to motion pictures. In examining the different stages of this industry, the central focus is the major motion-picture distributor because once a film has been leased from a producer, the distributor owns the exclusive rights to distribute the film to all the subsidiary program markets and maximizes profits by establishing the optimal sequencing, length of run and time clearance between these exhibition "windows."[2] The videocassette window has been moved to the front of the line, after theatrical release, due to the high net revenues generated when consumers buy cassettes directly or retailers acquire an inventory for rentals.

The question of increasing concentration is explored through the Industrial Organization Model (Caves, 1987; Scherer, 1980), which exemplifies the structure-conduct-performance paradigm of applied microeconomic analysis.

BRIEF HISTORY OF PRE-RECORDED VIDEOCASSETTE INDUSTRY

The main reason for the popularity of the VCR can be attributed to its versatility: immediate time-shifting, playback of home video photography made by portable video cameras, and playback of purchased or rented prerecorded videocassettes. Time-shifting was originally the prime motive for VCR ownership. This function provides the consumer with a significant gain in flexibility over when programs can be viewed.

In addition, many VCRs have been purchased for watching movies at home. As the VCR penetration dramatically increased this decade, the sale and rental of prerecorded videocassettes became an industry of its own, and VCR use for this purpose now overshadows time-shifting by a wide margin. The growth of the PRVC industry nearly tripled in size the last 3 years, growing from some 50 million annual sales to nearly 150 in 1989 (*Video Week,* 1983–1989).

The industry really started in 1978 when Andre Blay persuaded Twentieth-Century-Fox to sell him the cassette production rights to 50 movies for $6,000 each. Initially, very few cassettes were sold. Due to the high

[2]According to Waterman (1985), distributors determine optimal sequencing by employing price discrimination, that is, choosing the succeeding window according to the next highest contribution to net revenues.

price, however, some retailers began to rent cassettes rather than selling them, and the rental business was born ("The video revolution," 1985).

Since its inception, the prerecorded video program market has passed through two major development stages and is now embarked on a third one. The first stage was the *development* of the industry, heavily reliant on adult programming. The second stage was the *expansion* and legitimization of the industry, based on Hollywood feature film programming. The third stage, currently underway, represents the further expansion, *maturation* and *diversification* of the industry based on nontheatrical movies and nonadult programming, such as children's material, instructional, and music videos. This stage materialized when large-scale sales demonstrated that profitable submarkets also could be served with distribution patterns best suited to these genres (O'Donnell, 1985). The number of titles marketed in these genres continues to increase, even though theatrical movie videos still dominate the current retail market.

One of the factors that spawned this development is that it took the movie studios only 4 years to release almost all the films made in the past 50 years in this new entertainment format. Obviously, the Hollywood majors have depleted their movie archives, and new theatricals and made-for-VCR movies will have to be produced to fill the gap between supply and demand (Sunshine, 1985). The PRVC industry now generates over $10 billion from consumer rentals and sales combined (Bierbaum, 1989). For the Hollywood majors, the home video market now provides the largest source of film revenues after theatrical box office receipts. It would eclipse box office revenues if the distributors were able to share in revenues accruing from consumer rentals which account for at least 70% of the total consumer market (Bierbaum, 1989).

ECONOMICS OF THE DIFFERENT LEVELS OF THE INDUSTRY

The structure of the PRVC industry is unique because products go through more hands than in the traditional mass media industry such as motion pictures, broadcasting, or cable. These industries usually have three vertical stages—production, distribution, and exhibition; whereas the PRVC industry has five different levels—production, duplication, distribution, wholesale, and retail (Waterman, 1985). These extra layers are attributable to the fact that VCRs are marketed like magazines and paperback books and require individual copies and retail (and wholesale) outlets for direct consumer purchase.

Each level of the industry is examined, with particular focus on the degree of concentration and existence or absence of vertical integration. The special relationship between the production and distribution levels must be explored first. These two levels have historically been linked together because all Hollywood major producers have their own distribution arms. Given such widespread vertical integration, the economic analysis of both production and distribution levels is combined.

Production/Distribution Stage

According to *Video Week* (1988), feature films accounted for an average of 79.5% of all VCR revenues for the period of 1984–1988. Because the major film studios dominate the motion picture industry, this control will be directly transferred to VCRs as they distribute their archives of movies. Independent PRVC producers have also sought entry into this market, and lacking distribution facilities, they often have linked up with the majors. The relationship between the Hollywood majors and the independents has been one of the financier and the financed. Yet, no long-term contracts exist, which sometimes is the case at the duplication level. The majors decide to finance (or to purchase distribution rights to) a particular independent's film production solely on the basis of the perceived marketability or profitability of the former. This process is exactly the same as in the film industry. However, independent distributors in the PRVC industry more aggressively search for products to fill up the distribution network by purchasing rights for either theatrical or made-for-video films from independent producers. Some partially integrated independent producers/distributors, such as Vestron, have distributed up to several dozen made-for-video titles a year ("U.S. video independents," 1986) and are now considered major players at the distribution level. Table 2.1 indicates the trend in market shares of cassette sales since 1983. It is very clear that the major motion-picture distributors, either singly or in joint venture with others, dominate this segment of the industry; they account for nearly two thirds of the industry sales each year.

Two measures of market concentration, the four-film concentration ratio and the Herfindahl-Hirschman Index,[3] demonstrate a gradual downward drift in concentration between 1983 and 1988, with a stabilization since 1985. According to Department of Justice Antitrust "Guide-

[3]The Herfindahl-Hirschman Index is calculated by adding the squared shares of all firms in an industry. The H-H Index for 1985 was calculated with the assumption that there are five firms with .8% share each under the "All others" category in Table 2.1. Similar assumptions were made for other years.

TABLE 2.1
Market Shares for VCR Distributors
(Calculated From Annual $ Sales)

Distributor	*1988*	*1987*	*1986*	*1985*	*1984*[a]	*1983*[a]
CBS/Fox	9.2	10.0%	13.7%	13.5%	18.6%	18.2%
Vestron	6.1	6.4	8.0	10.4	10.0	6.0
RCA/Columbia	7.4	7.8	9.2	10.1	8.5	12.9
MGM/UA	4.5	4.3	7.5	9.0	6.5	10.0
Warner	9.8	9.1	8.5	8.3	9.0	10.0
Paramount	10.7	12.8	11.0	8.3	10.5	12.0
MCA	11.1	5.3	7.1	7.0	8.5	8.1
Disney	11.4	8.9	8.2	6.2	6.7	5.3
Thorn	n.a.	n.a.	n.a.	6.2	5.0	5.0
Embassy	n.a.	n.a.	3.4	3.6	4.5	4.1
Media	3.3	5.5	3.4	3.2	1.8	—
New World	2.2	2.2	2.4	2.4	—	—
Karl-Lorimar[b]	2.2	4.7	3.4	2.4	2.1	—
IVE	2.5	1.4	1.8	2.4	1.2	—
Prism	n.a.	1.0	1.2	1.2	1.1	—
Western	n.a.	n.a.	n.a.	1.0	—	—
HBO/Cannon (HBO)	4.4	6.6	6.4	6.2	n.a.	n.a.
Nelson	2.4	3.7	—	—	—	—
Cinema Group	n.a.	1.0	—	—	—	—
Orion	2.8	—	—	—	—	—
Goodtimes	2.0	—	—	—	—	—
Virgin	1.4	—	—	—	—	—
Video Treasures	1.0	—	—	—	—	—
All Others	5.6	9.3	4.8	—	6.0	8.4
CR_4*	43.0	40.8	42.4	43.0	48.1	53.1
H-H Index**	.071	.066	.087	.078	.092	.101

[a]Some data are missing, unreported or not available (n.a.)
[b]Lorimar has separated from Karl and was merged in late 1988 with Warner.
*The four-firm concentration ratio is the combined market shares of the top four firms.
**The Herfindahl-Hirschman Index is the sum of the squared market shares for each firm.
Source: Courtesy of *Video Week*, Annual Editions, 1983–1989.

lines," these indices indicate a transformation from a moderately concentrated market structure in the first few years to a monopolistically competitive one today. This is largely because rapid industry growth and maturation has opened up avenues for nontraditional distributors to discover a product niche. For example, Vestron, which did not exist in 1981, is shown to be a significant factor in diluting the market concentration by the major film studios. Also, the fact that small independents, such as Media Home Entertainment, more than tripled its sales to $54 million, and International Video Entertainment (IVE) almost quadruped its sales to $40 million in 1 year (from 1984 to 1985) is another indication of the transformation of the industry (*Video Week,* 1986).

Thus, although the role of the independent distributor is quite well defined in this segment of the industry, this does not mean that independents have smooth sailing. Mergers, acquisitions, and bankruptcies abound among the independents, and with the possible exception of small retail outlets, they clearly are the least stable segment of the entire VCR industry (Seideman, 1986). The independents' increasing financial difficulties are partially a result of the PRVC market maturity. First, consumers' tastes for rental videos have been refined over the past several years, and they are now more demanding in terms of PRVC quality. This is born out by the fact that the number of second-hand prerecorded videocassettes utilized by start-up rental outlets to build a quick inventory has decreased drastically in recent years. Another indication of viewers' demand for quality is the fact that video retail stores are not as willing as before to display "B" and "C" movie videos on their shelves. Hence, there seems to be a growing fixation on "A" quality movies, which is understandable because they generate a disproportionately large share of video revenues (Bierbaum, 1989).

The "B" movie label, as opposed to the mainstream, high-budget "A" movie label, usually refers to substream films that are especially favored by minority audiences (e.g., cult, extreme violence, high-brow, foreign, adult). The "C" movie label frequently refers to films that are considered to be commercial failures, and often means low-budget films (Hellman & Soramaki, 1985).

The evidence indicates that independent producers/distributors have made inroads into the PRVC market dominated by the Hollywood majors by appealing to specialty "B" and "C" movies and "how-to" or special interest video programming, which is examined in depth later. In this later category, independents actually created new demand for certain PRVCs, such as "Jane Fonda's Workout," and "Jack Nicklaus' Golf My Way," instead of imitating the major companies' products. These facts seem to support Hellman and Soramaki's (1985) contention that "the small- and medium-sized companies provide new material to be picked up and modified by the major producers" (p. 125). Regrettably, an ominous trend toward reconcentrating the industry may be soon approaching because of financial problems that Vestron has had in producing motion pictures and the recent merger activity by Warner Communications. In late 1988, Warner acquired Lorimar, including its home video division, and in mid-1989 found itself embroiled in a merger with Time, Inc., whose HBO division is a major home video distributor. With this merger now in place, this has created the largest

VCR distributor (HBO-Warner-Lorimar) with 16.4% of the current market (see Table 2.1).

Duplication Stage

Although there are quite a few small duplication facilities nationwide,[4] the majority of prerecorded cassette duplication for the home video market has been achieved by three major companies: VCA/Technicolor Duplicating Corporation, the industry leader with a market share averaging about 40% (recently merged with Technicolor Videocassette Inc. and a subsidiary of Video Corporation of America); Bell & Howell/Columbia/Paramount, the second largest firm with a market share averaging about 30%; and CBS/Fox Video, the third largest company with about 15% of the duplicating market (*Video Week*, 1986). By mid-1987, CBS/Fox exited the duplication sphere by selling its facilities to the VCA/Technicolor and giving this firm roughly half of the total industry sales. A condition of the sale was that CBS/Fox will "continue to use the duplicating facilities on the same basis and at the same prices under the new management" ("Technicolor gets CBS/Fox," 1987).

Several smaller national companies, most notably, IVE/Creative and Media, have been around since the birth of this industry with average market shares of 4% to 5%. A relative newcomer, GTK, entered in 1986 and has captured a slightly higher market share than these other two firms. By all measures of market concentration, this is an extremely concentrated industry and undoubtedly will remain so unless the smaller companies can somehow lure away some of the business of the major distributors.

One reason accounting for this high concentration seems to be the high cost of the duplicating equipment, which does not become appreciably lower with volume. Furthermore, cassettes are still mostly duplicated in "real time," which means that a 2-hour tape will take 2 hours to reproduce—an expensive and time-consuming process (Fairfield Group, 1984; Waterman, 1985). In recent years, some companies have started to use high-speed video duplicators that can copy a 2-hour cassette in less than 1 minute in the hope that the unit cost of duplication would go down significantly. With duplicating being highly labor-intensive (changing from one cassette to another to be copied) however, this technology

---

[4]Illegal duplication by small operators has posed periodic problems for major titles, but with stricter copyright enforcement and some more sophisticated antitheft protection, such practices account for less than 10% of the total volume (Bierbaum, 1989).

has not lived up to the high expectation of those duplicators ("Sell-through," 1987).

Although industrial and educational duplication accounts for 30% to 60% of the revenue of the three largest duplicators, many such duplicators rely on their movie customers, primarily the Hollywood majors, for the continual flow of orders to pay for their overhead. Some of the major duplicators, therefore, offer discounts to the Hollywood majors when the latter order their 35mm theatrical film processing and video duplicating together (Fairfield Group, 1984).

Paramount, Columbia, and CBS/Fox (until mid-1987) had been vertically integrated through ownership. At least for the past several years, other Hollywood majors (Warner, MCA, MGM/UA) and even Disney and Vestron have used the same duplicators (see Fig. 2.1). Thus, this constitutes vertical integration via long-term contract. In terms of control,

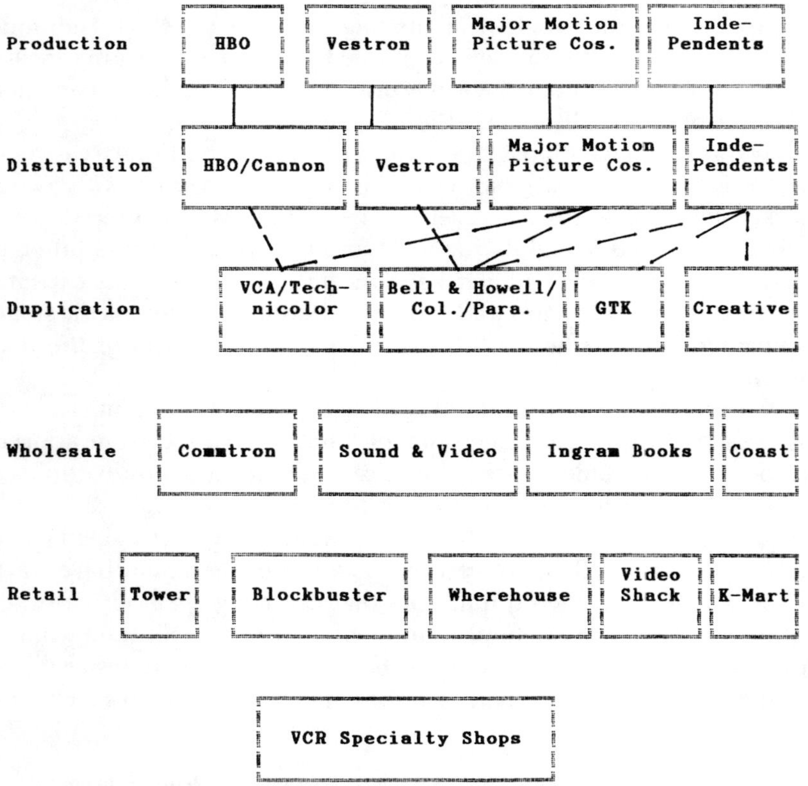

FIG. 2.1. Vertical integration in the PRVC industry. (*Solid lines indicate VI by ownership; dotted lines by exclusive contract.)

what is the difference between a long-term contract with an independent company and a long-term contract with a subsidiary? Although vertical integration through ownership is advantageous in internalizing decisions and negotiations and facilitating the long-range planning of the organization, it often is possible to achieve the same results through vertical integration by long-term contract with an independent company.

Litman pointed out that the television networks had achieved and solidified their triopolistic control through very restrictive affiliation contracts that gave the network an option over its affiliates' broadcast hours and reduced the ability of the station to reject network programs (Litman & Eun, 1981). It might be the case that the same type of contracts are now prevalent between the Hollywood majors and videocassette duplicators. As long as the quality of the product is kept high, vertical integration by contract may be as powerful a tool for achieving vertical control as vertical integration through ownership and involve less capital expenditures and risk.

Wholesale Stage

Wholesale distribution of prerecorded videocassettes is handled exclusively by independent video specialty wholesalers. Thus, there is no vertical integration by the major distributors, at least through the ownership mechanism. One thing to note, however, is the fact that sometimes the major distributors bypass the wholesale level and sell directly to the retail stores. This bypass is common when distributors sell their products to mass merchandisers and large chain stores rather than small video specialty shops because large orders yield quantity discounts for each side (Waterman, 1985). For this reason, direct distribution of PRVCs to retailers is still quite limited and does not divert or threaten independent wholesalers to a significant degree. Furthermore, despite longstanding affiliations between many of the Hollywood majors and record companies, the financial troubles of the latter made film studios wary of using the record companies as distribution arms (although the latter are used for distribution of music videos). This decision was reinforced by the fact that independent video specialty wholesalers were and are willing to warehouse large product inventory and thereby assume the associated risks of oversupply.

The four largest national video wholesalers are Sound & Video Unlimited, Commtron, Ingram Books, and Coast Video, and at least 20 national wholesalers are said to operate across the country (Fairfield Group, 1984). Figures for their individual market shares are not available, but it is known that some of these wholesalers tend to specialize with certain distributors. Although such a large number of wholesalers implies a competitive industry, this may not be the case in smaller communities that are only served by one or two regional wholesalers located in nearby big cities.

Retail Stage

Over the years, the number of videocassette retailers has increased to around 25,000, which indicates a rather competitive sector of the industry (Mayer & Sweeting, 1986). This is also an indication of the ease of entry. There is no vertical control extended to this level by the Hollywood majors and only a single tie by one of the smaller independent distributors (IVE).

Because video rentals tend to draw customers from a small (perhaps 2–3 mile) radius, some of the "mom-and-pop" shops have been able to coexist so far with large home entertainment chains like Wherehouse Entertainment and Tower Records as well as nationwide operations, such as Blockbuster and National Video. However, some shakeouts have occurred, especially among small-scale video specialty stores. The entry of mass merchandisers, such as K-mart and Sears and the emergence of a network of computerized vending machines that provide rented prerecorded tapes to credit card holders 24 hours a day, might have escalated this trend, even though their negative impact has been estimated to be rather minimal (Bednarski, 1985). Three to four thousand retailers are said to go bankrupt every year. Naturally, the number of new small video specialty stores has decreased. In general, it is safe to assume that the retail market is now somewhat more concentrated than before. But this is still the easiest submarket to enter among five different levels in the PRVC industry, judging from the large number of stores that sell pre-recorded videocassettes and the fact that videotapes are ubiquitously available in convenience stores and supermarkets much like popular magazines and paperbacks.

POTENTIAL FOR ANTI-COMPETITIVE BEHAVIOR

Bottlenecks for potential entrants to this industry seem to be in the production, distribution, and duplication levels where the Hollywood majors dominate and where economies of scale prevail to one extent or another. Is multiple stage entry absolutely critical now? Evidence suggests that there are numerous independents at all three levels as well as at the wholesale and retail levels (see Fig. 2.1). This may be because the major distributors sometimes need to purchase titles from independent producers; thus, they have no incentives to drive all independents out of business.

Another reason may be that the Hollywood majors, acting in isolation have insufficient market shares in any of these three levels to exercise their market leverage to drive out the independents, and they have not yet been able to forge a spirit of cooperation to act in concert.

Another anti-competitive factor takes the form of publicity. Among different genres of prerecorded videocassettes, theatrical movie videos and music videos enjoy greater name recognition because of their previous screening in movie theaters and broadcast and cable (MTV) television, respectively. Moreover, the product does not always have to be critically acclaimed. In the case of theatrical films, well-publicized and expensive box-office flops, if properly handled, often do quite well in the video stores. Yet, instances of unpublicized cult films with no theatrical track record no longer can count on the video market to return their investment (Kips, 1987).

From a competitive standpoint, videocassette distributors of those two genres (theatrical movies and music videos) benefit from not having to spend vast additional amounts on advertising to promote their video products because movies are advertised extensively when they are first theatrically released, and music videos get enough exposure through music video programs on television. Yet, a very recent trend shows major companies bolstering their initial advantage by spending significant amount of new money on their blockbuster video releases and establishing merchandise tie-ins with video product advertising ("MGM, P&G off," 1989). The "pulling power" of large national advertising budgets demonstrates the formidable advantage of media conglomerates to be able to take advantage of spillover effects. It is no wonder, then, that some of the independent video distributors are struggling to get public awareness and shelf space in bookstores to handle sales and promotional displays of their videos. The main reason for refusal is lack of room or perceived risks of diversifying too much by handling two distinct media.

PRODUCT DIFFERENTIATION AS A DECONCENTRATING FORCE

As explained, the major distributors utilize product differentiation (through advertising) as a barrier to entry by creating a perceived quality difference between their own movies (brands) and those of their competitors. But, another form of differentiation also is quite prevalent—*differen-*

tiation by product type. Because the bulk of currently available tapes are theatrical movies produced by the Hollywood majors, it is sometimes difficult for independent video companies to compete with those studios in the traditional category of broad appeal movies. Realizing this, many firms have been successful at producing and marketing other genres of videocassettes, discussed here, that have a distinct product nature.

Public-Domain Classic Films

A specialized market in public-domain classic films has been cultivated by a group of independent distributors, most notably, Kartes Video and Crown Video. These films can be duplicated without paying any royalties to the original studios. In the 1930s and 1940s, many studios allowed their film libraries to go out of copyright, wrongfully thinking that there would never be a market for classic movies (Sunshine, 1985). This made product differentiation possible for several firms because the retail price of those classic films ($7 to $20 each), is substantially lower than recently released films by the Hollywood majors. However, the number of public-domain films are limited, very unlikely to grow in the future, and this genre has reached a saturation point where various suppliers market the same title and compete solely on price (Lilienthal, 1985).

Pornography

From the inception of the prerecorded videocassette industry, adult-oriented movies have been a substantial component of sales and rentals. This submarket is composed almost exclusively of independent producers/distributors. Although data are difficult to obtain because of social stigmas, various trade sources now place the adult video share of the market at 10% to 15% (National Association of Broadcasters, 1989). This is roughly one third of its market share in the early years when it was a driving force helping to establish the industry (*Video Week,* 1983–1989). Entry into this submarket is relatively easy for new producers because of the new technology of the camcorder, modest product budgets, low production values, and dubious scripts. According to one trade source, the number of adult titles reaching the market each year has risen from about 400 to 1,700, and the films are being made with more sophisticated plots and more romance to appeal to new audiences, particularly women, who would not go to X-rated movie theaters but would watch hardcore cassettes at home (*San Francisco Chronicle,* 1984).

Children's Videos

After theatrical films, children's videos may constitute the most important home video programming category. It is estimated that children's videos accounted for over one half of the prerecorded, *non-theatrical* video rental market over the period of 1984–1988, and a steady 8% to 10% of the theatrical market (*Video Week,* 1983–1989). Recognizing this trend, it no longer is unusual for major suppliers to have a children's video division devoted to acquiring, developing, and marketing products in this genre ("Tickling the children," 1985). As indicated in Table 2.1, Walt Disney is the seventh biggest producer in the prerecorded videocassette industry as a whole. Thus, there is no doubt that it is a dominating force in this submarket. Other suppliers of children's videos include such independents as Vestron, Family Home Entertainment, and even some of the Hollywood majors. Children's videos like *Cinderella* and *Lady and the Tramp* have ranked among the leading sell-through tapes during the last several years.

Videocassettes makes it easier to monitor and control youngster's television viewing since children have no option of changing channels. Suppliers and home video retailers have learned that children's videos perform best as a sell-through item because children are fond of watching favorite features and cartoon characters repeatedly—a key characteristic of the sell-through market (see Jordan, chapter 9 and Dobrow, chapter 10, this volume).

Music Videos

The release of *The Making of Michael Jackson's "Thriller"* by Vestron in late 1983 created a tidal wave of interest in this genre among the music record companies. And when coupled with the earlier popularity of MTV among teens and young adults, it clearly demonstrated the potential sales levels of music video as a category distinct from movies, although it simultaneously misled some to assume that all music videos would sell equally as well (O'Donnell, 1985).

There are some other factors responsible for spurring the increased production of music videos. More and more VCRs and prerecorded videocassettes are being sold with high-fidelity/stereo capability. Similarly, with Korean entry into the VCR market in 1985, even lower priced VCRs appeared, thus bringing the price level well within the purchasing power of the average teenager. On the technical side, the duplication costs for manufacturing music videocassettes are steadily decreasing (O'Donnell, 1985).

In this submarket, the major record companies are the most powerful forces. Most of them distribute their videos to record retailing outlets because this product more closely resembles records than standard VCRs. The magnitude of their operations, however, is relatively small, accounting for only 2% to 3% of the PRVC market in the last few years and hence not much of a competitive threat. Music video also is a category that lends itself as a sales item (even more so than children's videos) because consumers play them repeatedly. Research by Sony has shown that its "Video 45 cassette" (typically 4 songs, 15–20 minutes long) are viewed an average of 14 times, with over 25% of those surveyed saying they have screened the tapes over 20 times (O'Donnell, 1985).

"How-To" Videos

Interest in the area of instructional video was heightened by the incredible success of the breakthrough "how-to" videocassette, *Jane Fonda's Workout,* released in April 1982 by Karl Video. *Workout* set sales records for several consecutive years until the majors decided to offer very attractive "sell-through" prices on blockbuster theatricals such as *ET* (14.5 million units), *Cinderella* (7.9 million units) and *Top Gun* (2.85 million). *Workout's* 1 million worldwide sales now seem inconsequential in comparison to the major league sales of the big studios; yet *Workout* was truly a pioneer video that spawned a cottage industry of imitations, with over 60 new exercise video titles released in the past several years, including Fonda's new *Startup* video with 800,000 unit sales (*Video Week,* 1988).

The success of Fonda's product clearly demonstrates that a star can be a key element in selling a "how-to" videocassette. Some other ingredients for success of this genre are repeatability, a lower retail price, interaction with viewers, and a subject matter that is taught better on videocassettes than in any other media (Sunshine, 1985). Yet, beneath all the fanfare and publicity, this sector has never reached a market share of more than 5% to 6%, and its lustre has faded because some companies, such as CBS/Fox, Simon and Schuster, Lorimar, and Caravatt Communications have either withdrawn from this submarket altogether or scaled down their operations due to low profitability ("U.S. special interest," 1988).

COMPETITION WITH OTHER HOME VIDEO ENTERTAINMENT MEDIA

As far as prerecorded videocassettes are concerned, nothing seems to threaten their existence, except competition from pay-per-view (PPV) cable. PPV requires an "addressable converter" that can be authorized to receive a special transmission from the cable offices in response to a phone order from a person's home. As of early 1989, there were estimated to be 18.9 million addressable cable households, a 20.9% penetration rate ("At the crossroads," 1988). This number has jumped considerably in the last few years as new cable systems and upgrades of existing systems are featuring this new technology. Unfortunately, with no groundswell of consumer interest, and at an additional per household expenditure of $150 to $200, most cable operators have hesitated to obsolesce their current converters in favor of addressable ones ("At the crossroads," 1988).

The leading PPV national networks are Viewer's Choice (with 4.6 million subscribers) and Request Television (with 3.3 million subscribers). Other networks, Cable Video Store and Playboy on Demand, have at least 1 million subscribers each ("Pay per view," 1989). However, the industry is still limping along, and none of the networks has made a profit. This is due to the relatively low national penetration rate and infrequent usage of the service from those who do subscribe. The biggest financial successes on PPV have been for special event *live* programming such as championship prize fights, *Wrestlemania,* and musical concerts. These events usually are premium priced because of their special nature and because they do not lend themselves to repeat offerings, they are unique and face no direct competition from the VCR industry.

For the normal diet of theatrical movies, PPV networks claim that this technology is superior for obtaining a favorite program than going to a video rental store. In fact, videocassette rental necessitates two trips to the stores. Despite this disadvantage, however, a recent survey indicates that viewers prefer to see a particular movie on a rental-video basis, even if it were released simultaneously on PPV cable and at the same price (Electronic Media/SRI Survey, 1987). In reality, prerecorded videocassettes have a strong advantage over PPV because it still remains cheaper to rent a tape ($2 to $3) than to subscribe to a PPV program ($4 to $5), and the viewer has additional rewind and playback flexibility with the VCR. The future success of PPV will depend on the speed in which it can penetrate existing cable systems and whether it obtains a preferred "release window" over VCRs.

First Sale Doctrine

One negative element for the PRVC industry is the fact that Hollywood studios are more enthusiastic about PPV than prerecorded videocassettes. This concerns the "First Sale Doctrine" of the Copyright Act. Under this doctrine, distributors lose all control of their prerecorded tapes once that tape is wholesaled to a retailer. Without having to pay any additional royalties, retailers have been able to keep the sizable revenues generated from rentals for themselves. Because of this situation, VCR distributors historically have been forced to set too high of a price for wholesale cassettes, thereby ruining the direct consumer sell-through market and stimulating rentals (Waterman, 1985). This is quite different from the distribution process for pay-cable and PPV where cable networks purchase movie rights based on the number of their subscribers and then split the consumer revenues with the cable systems (e.g., the motion picture theater model) according to a specified formula. Distributors perceive PPV to be more favorable to them in the future and have tried to enthusiastically promote this technology. It is in that respect that PPV cable is the most formidable competitor to the prerecorded videocassette.

The major distributors, however, have not always responded favorably to the calls from PPV programmers and cable operators to provide an earlier "release window" for PPV. The distributors have released some films to PPV 6 weeks earlier than to the video, but not on a permanent basis. Distributors are cautious because, at present, PPV has relatively low penetration rates and prerecorded video sales generate as much revenues as the box office. The latest evidence suggests that VCRs are winning this crucial battle and regaining their favored status (Bierbaum, 1989).

The distributors' push toward PPV will definitely diminish if the relative balance shifts toward more consumer purchases and less rentals or if distributors can somehow share in the rental revenues. Both situations are actually happening to some degree now.

Rental Versus Sell-Through

Historically, VCR rental revenues have exceeded sales revenues by a factor of 4 to 1 (Bierbaum, 1989; *Video Week,* 1988). Because distributors have a greater profit interest in the sell-through market, they have made conscious efforts in recent years to stimulate sales through the practice of price discrimination, commonly referred to as *price polarization.*

New video releases of hit movies will have suggested retail prices of $89.95. (Wholesale prices are naturally lower.) This will dissuade most consumer purchases, but retailers will purchase sufficient "depth" to handle expected rental demand. This high price period will last for several months until the video has circulated among the rental crowd. The distributor will then lower its wholesale and suggested retail prices by $30 or more to stimulate the sell-through market. At some later point in time, the video will be marked down for special holiday promotions to the $20 to $30 range. This favorable price will naturally entice many consumers into "library building" (see Heintz, chapter 8, this volume). The enormous unit sales of *ET, Top Gun,* and other films on videocassette is directly attributable to such a pricing strategy although the Fall 1989 release of *Batman* skipped the high price period and went directly to the holiday price of $24.95 or less.

Videocassette sales are enormously profitable to the distributor; obtaining $5–$10 of net income from consumers per cassette is more than the amount the movie studios can obtain for a film shown on all other electronic media and theaters combined (Waterman, 1985). Owing to the distributors' price-tiering strategy, the average retail price of prerecorded videotapes has fallen steadily in recent years from $77 in 1982 to about one third that amount today (*Video Week,* 1983–1989) and naturally, unit sales have correspondingly skyrocketed, thereby reinforcing industry claims of a price-elastic product. In fact, in 1989 the direct sell-through market is now a $3 billion industry and when combined with over $7 billion from cassette rentals, it exceeds the revenues of the commercial TV networks. It is uncertain how much lower consumer prices can be expected to fall because this product passes through many vertical stages, each adding "value" to the product. Some recent evidence does indicate that prices in the range of $15 to $20 are becoming more prevalent in the industry with $14.95 expected to become the equilibrium selling point in the near future. One thing is certain. The specialized video stores are going to have to open up greater shelf space for sell-throughs to accommodate this growing segment of the industry. They cannot permit the mass merchandisers and video mail clubs to have this business all to themselves, even though the current profit margin is not as high as in renting tapes.

Pay-Per-Transaction

The price discrimination plan has caused problems for the video dealers in the rental side of their business. Because they have to pay high wholesale prices for videocassettes, they cannot buy as many copies of popular videos as they would like (the "depth of copy" problem). Whereas the average retailer used to buy a blockbuster video title for rental for every 100 VCR homes a few years ago, the equivalent figure for 1987 was every 200 VCR homes per copy ("National video wants," 1987). To respond to this problem, one large video store chain, National Video, devised a so-called "pay-per-transaction" (PPT) method. With PPT, distributors can obtain a higher stream of revenues from each video sold or leased directly to retailers and hence need not charge as high a wholesale price. To participate in this scheme, retailers usually pay a flat fee to obtain the right to rent out a cassette from a distributor and subsequently share between 30% to 50% of the rental fees with the distributor ("Behrens, 1986; "National video wants," 1987). Although at first this scheme elicited widespread industry participation from retailers and distributors alike, it has not proven to be a roaring success and its future remains in doubt. It has been plagued by electronic (and accounting) problems in keeping track of all the transactions and mistrust between the retailers and distributors. Only one major company retailer, Rentrak, and one distributor, Orion, still seems wedded to this approach.

CONCLUSION

This chapter has highlighted the market structure of the prerecorded videocassette industry. It was seen that video product is multifaceted in content, ranging from mass-appeal motion pictures, to pornography, children's video, and even the most esoteric "how to" instructional tapes. This product travels through many stages from its creative beginnings to its ultimate destination as a consumer retail item. Although each of these stages is large enough to be independent and competitive, vertical integration and concentration of control accompanies the production/distribution sector, where the large movie studios have transplanted their theatrical market power. Surprisingly, vertical integration has not tied together all of the stages of product as it did in the historical precedent of the motion picture industry, prior to the *Paramount* decrees. Small independents can still coexist with much larger players at each and every stage to one degree or another.

Critical questions to the future of the industry are what place in line will it stand as a release window for theatrical movies? And should it seek a closer balance between sales and rentals? The former question will depend on whether electronic home video competitors such as PPV or the newest technology of interactive video on demand (delivered by fiber optic cable) become widely available and generate efficiencies for distributors in establishing a premiere showcase for newly released theatricals.

As for the second issue, this really deals with the distribution of wealth between the retail and distribution sectors. Because of the "first sale doctrine," video retailers have not needed to countervail the power of the major distributors with such trade practices as "product splits,"[5] and "anti-blind bidding"[6] which are typical in the film exhibition sector. As sell-throughs become relatively cheaper in future years, this industry may develop like the book industry at the turn of the century which was originally a rental business and then entered a complete transformation.

Above all else, the prerecorded videocassette industry has launched a consumer revolution in program sovereignty both in terms of availability of diverse programming and freedom from the "temporal tyranny" of the program executives.

REFERENCES

At the crossroads. (1988, December). *Channels, 1989 Field Guide to the Electronic Environment,* p. 102.

Bednarski, P.J. (1985, November 11). Videocassette vendor setting up 7-Eleven network. *Electronic Media,* p. 33.

Behrens, S. (1986, September). Ron Berger's heresy. *Channels,* pp. 44–47.

Bierbaum, T. (1989, January 11). Year of growth for homevideo. *Variety,* p. 85.

Caves, R. (1987). *American industry: Structure, conduct, performance* (6th ed.). Englewood Cliffs, NJ: Prentice Hall.

Fairfield Group. (1984). *Home video cassette disc publishing: The dynamics of distribution, 1984 and beyond.* White Plains, NY: Knowledge Industries.

Hellman, H., & Soramaki, M. (1985). Economic concentration in the videocassette industry. *Journal of Communication, 37*(3), 122–34.

Kips, C. (1987, October 7). Homevideo now after TV oldies. *Variety,* pp. 1, 116.

Lilienthal, L. (1985, November). Public domain video sales slipping. *Billboard,* p. 3.

Litman, B., & Eun, S. (1981). The emerging oligopoly of pay-TV in the USA. *Telecommunications Policy, 5*(2), pp. 121–35.

Mayer, I., & Sweeting, P. (1986, January). 25,000 shops on Main Street. *Channels,* 1986 Field Guide, p. 76.

MGM, P&G off to see wizard: Biggest vid release ever. (1989, June 7). *Variety,* p. 39.

National Association of Broadcasters. (1989, April/May). Info-Pak.

National Video wants to see its revenue-share plan used widely. (1987, November). *Variety,* p. 32.

---

[5]A *product split* refers to a situation where an unwritten agreement is made among a distributor and local exhibitors so that all but one theater voluntarily refrain from seeking an exhibition license for a particular picture. This prevents or makes it difficult for exhibitors that are not part of the split to have access to quality first run films.

[6]*Blind bidding* refers to bids that exhibitors make without even seeing a particular film. Many exhibitors tolerated the blind bidding as a way of getting what they perceived to be scarce commodities, relatively high-budget films from Hollywood majors.

O'Donnell, J. (1985, January 16). Music video will prosper with growth of market for Made-For Product. *Variety,* p. 197.

Pay per view supply chart. (1989, March 13). *Cablevision,* p. 50.

Scherer, F.M. (1980). *Industrial market structure and economic performance* (2nd ed.). Chicago: Rand McNally.

Second VCR for one in seven U.S. households. (1987, September). *Screen Digest,* p. 200.

Seideman, T. (1986, July 12). Decline of B and C titles spurs indie vid shakeout. *Billboard,* p. 84.

Sell-through and video duplication. (1987, June). *Screen Digest,* pp. 134–5.

Staff. (1984, November 12). *San Francisco Chronicle,* p. 4.

Staff. (1986, July 9). *Television/Radio Age,* p. 108.

Staff. (1986, January 6). *Video Week,* pp. 1–5.

Staff. (1988, December 26). *Video Week,* pp. 1–5.

Sunshine, L. (1985, February 15). The videocassette business. *Publisher's Weekly,* pp. 36–53.

Technicolor gets CBS/Fox video copy plant. (1987, August). *Screen Digest,* p. 173.

The video revolution. (1985, August 6). *Newsweek,* p. 51.

Tickling the children's video marketplace. (1985, May 25). *Billboard,* p. KV2.

U.S. special interest video looks shaky. (1988, January). *Screen Digest,* p. 17.

U.S. video independents into film making. (1986, September). *Screen Digest,* p. 172.

Video Week. (1983–1989, January). Annual Industry Summaries.

Waterman, D. (1985). Prerecorded home video and the distribution of theatrical feature films. In E. Noam (Ed.), *Video media competition* (pp. 221–43). New York: Columbia University Press.

3

Audience Measurement in the VCR Environment: An Examination of Ratings Methodologies

Bruce C. Klopfenstein
Bowling Green State University

INTRODUCTION

The 1986–1987 television season marked the first time the A. C. Nielsen company found U.S. weekly television viewing to have *declined* from the previous year. The decline was even greater when measured by the new peoplemeter system (Kneale, 1988a). Another decline occurred in the 1987–1988 season (Nielsen, 1989) in spite of the fact that Nielsen also found multiple set households up to 59% of all television households, and more television and cable channels than ever available to U.S. television households ("Weekly TV viewing," 1988); almost 56% of all television households had cable television by early 1989 according to Nielsen ("May penetration," 1989, p. 45). Although the growth in video options has matured in the United States, it is worth noting that a similar change is now taking place in Europe (Buck, 1988; Durand, 1988; Kavanaugh, 1989; Nayeri, 1988).

The decline of the three networks' viewing audience continues and has been well documented (e.g., Hadlock, 1988; Hey, 1987; Kidder, Peabody, & Co., 1988; Kneale, 1988a; Knight, 1989; Knobeisdorff, 1988; Lerner, 1988; Lieberman, 1987; Lippman, 1989; Zoglin, 1988). Many people—both inside and outside of the network television industry—attribute some of the decline in television viewing to the new video kid on the block—the VCR. Broadcasters have changed programming strategies based on the impact of the VCR on audience behavior (Gelman, 1987; Hickey, 1988; Mahler, 1987; Rosenthal, 1987). This has led to a change in the way advertisers are using television as an advertising medium (e.g., McSherry, 1988; "Survey," 1988; Walley, 1988). The new video environment is a challenge to commercial and academic researchers alike (Benson, 1988; Goodhardt, Ehrenberg, & Collins; 1987; Kavanaugh, 1989; Krugman, 1985; Reed, 1989; Sternberg, 1989; Webster, 1989a, 1989b). Clearly, more information is needed on VCR users and their behavior.[1]

Nielsen reported that the VCR had been adopted by nearly 65% of all U.S. television households by early 1989; San Francisco had the highest adoption rate of the major markets, with at least one VCR in three out of four television homes ("VCRs in 65% of homes," 1989). VCR penetration in Europe has not lagged far behind (Luyken, 1987). The VCR diffusion pattern in the United States means that about 70% of U.S. households will have VCRs by late 1990.[2] More people have VCRs than any other video technology except color television. The VCR has attained its lofty perch in American households more rapidly than any previous communication technology except monochrome television (Klopfenstein, 1989a). VCR diffusion is expected to continue rapidly with penetration nearing upper limits in 1990 among middle-aged householders with children (Swanson & Klopfenstein, 1987). These households are especially attractive to advertisers. The number of multiple VCR households is also rising. As of 1986, one estimate already had 14% of VCR households having more than one (Zahradnik, 1986), whereas Nielsen put the figure at 21% by late 1988 (Nielsen Media Research, 1988).

Although VCR diffusion has been quite impressive, the same cannot be said of VCR measurement yet. Like research on previous new media technologies, scholarly investigation tends to lag behind initial diffusion of the medium (Williams, Rice, & Rogers, 1988). The purpose of this chapter is to review the current commercial methods of audience measurement. Although scholars and research practitioners are familiar with survey research techniques generally, few outside the industry (and even inside the industry) are very familiar with the techniques used to produce ratings data. An extensive review of the literature indicates that other

---

[1] Indeed, much existing academic VCR research has relied on convenience and other non-random sampling techniques due to the exploratory nature of much of that research (Klopfenstein, 1989b).

[2] Nielsen VCR penetration figures are, perhaps, the most cited figures available; many scholarly VCR articles began by citing Nielsen VCR penetration figures to support the importance of their investigations. However, they have historically been conservative (Klopfenstein, 1989b). Although Nielsen put VCR penetration at about 50% in 1987, a survey by Frank Magid & Associates found it to be nearly 60% (*Broadcasting*, 1987, October 19, p. 7). The discrepancy is also probably due to different sampling techniques.

than Beville (1985), little research is readily available on these methodologies (Hurwitz, 1984). Webster and Wakshlag (1985) also summarized available methods for measuring exposure to television. The diary (and household meter for television) have been accepted as the audience research methods of choice for the past 30 years. After examining the conflicting VCR research agendas to date, this chapter seeks to answer several general questions.

1. What commercial audience research methods are in use today?
2. What are the relative strengths and weaknesses of those methods?
3. What methods are being developed for the future?
4. What are the various methods' implications for VCR measurement?

VCR AUDIENCE MEASUREMENT AGENDAS

Little *published* research that accounts for the VCR's impact on the broadcast audience (Klopfenstein, 1989b) is currently available. Levy (1988) has discovered that there are even academics who question the importance of VCR research in spite of the fact that much work has yet to be done on the social impact of the VCR (Levy 1987, 1988) Research reported publicly by commercial research firms is generally limited to static tallies of average weekly VCR use. As of 1989, Nielsen still only reported VCR *recording* behavior but not *playback* behavior (Luchter, 1988).[3] AGB's first report on VCR use was released in March 1988 (Paskowski, 1988) shortly before that firm gave up its bid to challenge Nielsen in the U.S. ratings business (Trachtenberg, 1988). AGB was only beginning to release analyses of its VCR data (Sims, 1989). Nielsen was expected to begin reporting playback of rented or purchased videos (as well as home-recorded material) in late 1989 (Bierbaum, 1988b; Teinowitz, 1989).

Like cable television before it, the VCR was initially welcomed by broadcasters as the new technology served to increase their reach to audiences that might otherwise miss their programs (Levy, 1980a, 1980b, 1981, 1983). The advent of the VCR remote control (which made avoiding recorded commercials much easier, e.g., Lewin, 1988) and the proliferation of video rental stores (which allow viewers to actively select movies of their choice rather than passively accept those of the broadcast or cable programmer) have made the VCR as much of a substitute as it might be a complement of it.

[3] Some industry observers believe this is, perhaps, in response to network pressure (Gay, 1987).

Thus, members of the broadcasting community today often point out what they see as the limited impact of the VCR on their audiences. For example, in a column in the trade publication *Television/Radio Age,* a television station executive notes that the average VCR household rents only 2.8 tapes per month, and broadcast television programs remained the subject of most VCR recording activity (Stitt, 1989).[4] As in any industry, advocates for the broadcast industry can be expected to portray the impacts of a potential competitor in the best possible light. Broadcasters have pointed to such research in the past ("Home tape study," 1985; Metzger, 1986; "Study finds little," 1989).

Like broadcasters, Nielsen vice president Paul Lindstrom (1989) questioned the extent of the VCR's impact on broadcast audiences.[5] The head of the Nielsen Home Video Index introduced a summary of Nielsen research on the VCR by downplaying the rhetoric that has surrounded the VCR and its television viewing impact. Lindstrom noted that the VCR penetrated households with heavy television viewers first. Because further VCR penetration could be expected to be into households with lighter television use, the greatest effect of the VCR has *already* been felt. Klopfenstein (1988), on the other hand, found that respondents from non-VCR households were heavier television viewers than those from VCR households.

Lindstrom further pointed out that the "average" number of recordings in VCR households decreased from 1984 to 1987. Such an observation of a statistical mean can ignore the differences between early and later adopters in VCR use (Klopfenstein, 1989b; a portion of this research is discussed here). It seems reasonable to suggest, for example, that although earlier adopters bought their VCRs primarily to time-shift programming, more recent adopters may have been attracted by the ready and inexpensive access to movies or other prerecorded material on video.[6]

Because VCR growth accelerated dramatically in 1986–1987, there

---

[4] Nielsen reports that networks lead in the percentage of *programs* recorded, and that 67% of all recordings in November 1988 were originated on network affiliates (Nielsen, 1989). This allows broadcast executives like Stitt (1989) to point out the apparent dominance of the network affiliates in VCR recording activity. If the base were made recorded *minutes* rather than *programs,* a different conclusion might be drawn. For example, if one were to record a 30-minute soap opera and a 2-hour (120 minute) movie on HBO, this would mean that 50% of the program recording activity originated on the network affiliate, whereas 80% of the recorded minutes originated on HBO.

[5] Beville (1986a, 1986b), too, found the impact of the VCR to be relatively limited through 1985. Indeed, most analysts see cable as a much greater "threat" to broadcasting than VCRs today.

[6] There is also evidence that more recent adopters have had a difficult time learning to use their VCRs to record programming. VCR designers have been working hard to make their machines more user friendly for recording. One example of this is making bar code input devices available for programming purposes.

was a great influx of early adopters who may have been more interested in the playback functions of the VCR.[7] Indeed, in his analysis of Nielsen VCR usage data from this time period, Sternberg (1989) reached similar conclusions to Klopfenstein (1989b). After examining differences in data supplied by Nielsen on VCR usage behavior versus length of VCR presence, Sternberg (1989) observed: "Clearly the impact of VCRs on television ratings as we enter the 1990s will be far greater than it is today" (p. 82). Access to the Nielsen data vaults would be necessary to clear up the contradictions.

Other reasons to be cautious about the VCR use data released by Nielsen involve the conservative forces that often exist to preserve a status quo in a huge industry like broadcasting (see Meehan, 1984). There are several apparent manifestations of this conservative force. First, Nielsen rapidly switched to the new peoplemeter system as a response to the threat originally posed by a new competitor, AGB, in 1987 (Ostroff, 1989). As is discussed in greater detail later, the old diary system favored the broadcast networks and affiliated stations over cable and independent stations[8] (Huff, 1988; "Meters rattle," 1988; "New ratings scale," 1988). Second, Nielsen has charted VCR usage in a manner quite friendly to broadcasters (and contrary to the best interests of advertisers).[9] In a procedure called *ascription,* through 1989 Nielsen simply included *recorded* television shows in the ratings *regardless* of whether the shows were ever replayed.[10] Not only might advertisers be paying for shows that may never be watched, they also might be paying for commercials that the viewer can easily avoid by "zipping" or fast-forwarding during replay (Cassata & Irwin, 1989; Smith, 1989). Third, AGB folded its television measurement service when none of the three networks would support it. AGB had announced very specific plans to measure VCR playback activity and was in the process of releasing such information just prior to suspending its service.

The broadcast industry has also sponsored a VCR research project

---

[7] Von Hippel (1986) noted the differences between expected uses of new products and those that are created by adopters. Rogers (1986) called this process reinvention.

[8] WEWS-TV salesperson Larry Olevitch has studied the differences between Arbitron ratings based on its local household meter and Nielsen ratings based on its local diaries in the Cleveland market. His comparisons confirm significant discrepancies in the two methodologies.

[9] Although both advertisers and broadcasters purchase ratings services from Nielsen, broadcasters are the primary source of revenues for Nielsen ratings overall. About 90% of Nielsen's local ratings revenues come from television stations (Philport, 1989).

[10] Nielsen reported that network mini-series are among the most recorded programs, and up to 1.6 ratings points (representing about 1.5 million homes) were added to mini-series' ratings ("VCRs play heavy role," 1988). This does not consider whether or not the programs are ever viewed.

that indicated a relatively limited impact on the audience. The three broadcast networks and the National Association of Broadcasters fund "COLTAM," the Committee on Local Television Audience Measurement. In a preliminary study conducted by SRI and widely cited by broadcasters (e.g., "Home tape study," 1985), Metzger (1986) reported VCR use as having little negative impact on broadcasters. Commercial avoidance figures that used the universe of *all* taped shows (including those with no commercials) were cited. Other figures included households in which the VCR was not used the previous day as part of the universe to report commercial avoidance behaviors. The larger the base of nonusers, the more limited the impact of the VCR on the broadcast audience appears to be. In an effort to increase validity, Metzger limited self-reports of usage behaviors to the previous day.

Kaplan (1985) took the advertisers' point of view in arguing for better measurements of commercial avoidance. Yorke and Kitchen's (1985) data from personal interviews dramatically contradict those found by Metzger (1986). Yorke and Kitchen and Metzger, however, used different methods to measure VCR usage behavior, and this point is critical. Because broadcasters clearly wish to portray their industry in the best possible light, their research efforts can be expected to employ methods that will do just that. These problems call out for more *unbiased* VCR research *and* analysis. Much of the debate is being clouded by two sides with vastly different agendas.

PRIMARY VCR RESEARCH EXAMPLE

To further examine the contentions made by both broadcasters and Nielsen about the limited impact of the VCR on broadcasting, Klopfenstein (1989b) looked at data from a random telephone survey of a medium-sized market involving 1,000 respondents (583 from VCR households). Among the questions addressed were:

1. How do various adopting VCR households differ demographically, in presence of other technologies and in television viewing?
2. What differences in VCR recording use and in cassette rental behaviors exist versus length of VCR presence?
3. What differences exist in reported commercial avoidance behavior versus length of VCR presence?
4. Are there indications of a "wearout" factor in VCR usage (is VCR use *lower* vs. length of VCR presence)?
5. What are the possible implications for future VCR use?

Respondents were asked how often (regularly, occasionally, rarely, or never) they view each of 11 television program genres including situation comedies, hour-long dramatic series, daytime soap operas, sports, local news, national news, news magazines, game shows, movies on broadcast television, late night talk shows, and public television shows. VCR owners were later asked how often they *record* each of these formats. A number of items were developed to get at respondent attitudes toward broadcast television and possible motivations for VCR adoption such as commercial avoidance, time-shifting, movie rental, cable use, and purposive television viewing (as indicated by use of television listings). Respondents were also asked about their recording and commercial avoidance behaviors.

Respondents from VCR households were generally found to view *less* television than those from non-VCR households. Direct correlations were found between length of VCR presence and household income, presence of children in the household, and presence of other communication technologies such as number of television sets, number of VCRs, video cameras, cable television, pay cable, personal computers, telephone answering machines, and compact disc players. The longer a household had a VCR, the more likely the respondent was to have used it for recording purposes. A direct correlation was found between length of VCR presence and recording of sports, news, and news magazine shows. Other attitude and belief items indicated a correlation in the same direction as seen in Table 3.1.

Earlier VCR adopters are also of higher socioeconomic status than more recent adopters. Earlier adopters learned the utility of using the VCR to record and reported higher frequencies of recording several program types. It is not clear if these types are more attractive for recording, and more research is needed in this area. There was no evidence in this study to indicate a "wearout" factor in VCR usage. If anything, VCR use may increase over time as the user learns both how to program it to record and the advantages in doing so. This behavior may be offset to at least some extent by a possible decline in VCR use if there is a "novelty effect" in recent adopters.

There was no relationship between length of VCR presence and self-reported hours of television viewing. This contradicts the contention that the most recent VCR adopters are *light* television viewers (a counter-intuitive notion because the VCR is only now penetrating lower income households where television viewing is generally high). There were also no relationships between length of VCR presence and reported viewing of the 11 genres of television shows. Klopfenstein (1988) did find differences between VCR and non-VCR households in 7 of the categories with respondents from *non*-VCR households reporting greater viewing of broadcast television. There was surprisingly little difference between

TABLE 3.1
Pearson Correlations Between Self-Reported Frequency of Recording Behaviors and Length of VCR Presence and by Income

Behavior	*N = 412* *All* *r value*	*N = 105* *<$30 K* *r value*	*N = 226* *≥ $30 K* *r value*
When you record a program off the air, how often do you . . .			
Record the program while watching it	.029	.069	.022
Record a program while watching a different program	.119**	.037	.142*
Record a program while not at home when the show airs	.128*	−.011	.182**
Record a program while sleeping when the show airs	.126**	.202*	.149*
Delete the commercials while recording programs	.099*	.153	.070
Fast-forward past commercials when playing back recorded programs	.028	−.008	.027
About how many days is your VCR used to record TV programs in an average week?	.041	.000	.074
Beliefs [strongly agree, agree, disagree, strongly disagree]			
Recording network television is an efficient way to use my time	.172***	.063	.154**
Avoiding commercials is an important reason to have a VCR	.010	.102	−.060
Renting tapes is the primary reason for having a VCR	−.050	.017	−.126*

*$p<.05$, **$p<.01$, ***$p<.001$.

Regularly was recoded as 4, occasionally as 3, rarely as 2 and never as 1. Length of VCR presence was coded < 1 year as 1, 1 to <2 years as 2, 2 to <3 years as 3, and 3 + years as 4. Income categories do not add up to 412 due to missing values.

Source: Kopfenstein (1989b)

recent and earlier VCR adopters in self-reported cassette rental behaviors although 63.5% of all VCR respondents agreed or strongly agreed that "renting tapes is the primary reason for having a VCR." It remains an open question whether VCR owners will become bored with renting tapes, although there was no evidence in this study to suggest that has happened or will happen.

Clearly, a telephone survey in which respondents may give socially desireable responses has its limitations (Henke & Donahue, 1989, is another recent telephone study). There are always validity questions with self-reports of behavior (Klopfenstein & Swanson, 1987). A longitudinal study of VCR households would be most useful, but also time consuming and costly. Until then, we will have to make due with the tools that are available to us. Although a telephone survey can get at respondent attitudes toward VCRs and their use, commercial research firms have invested millions of dollars into the various methods used to measure daily exposure to electronic media. These methods are now reviewed in light of their application to measurement of VCR usage behavior.

CURRENT COMMERCIAL AUDIENCE MEASUREMENT METHODS

Measuring the television audience today is more challenging than ever before (Webster, 1989a, 1989b). VCR measurement (both VCR use and its displacement of other media use) is only one element of the increasingly complex television audience measurement equation. Although Nielsen has the economic resources to invest in what may be the most valid method of measuring television viewing to date, the peoplemeter, academic researchers do not have the resources to gain access to and utilize this technology. Because the Nielsen data are proprietary, access to the data is also closed for the most part to nonsubscribers, including academics (Philport, personal communication, May 23, 1989). AGB had been making some data available for outside scrutiny (Ehrenberg & Wakshlag, 1987; Sims, 1989). This section of the chapter examines the strengths and weaknesses of the various methods used in commercial audience research—including the less expensive methods of radio research.

Similarities Between the Evolving Video Environment and the Mature Radio Environment

Radio has been a neglected subject of mass communication research study since the advent of television. Ironically, the new television/video environment with its multiplicity of program options seems analogous to the radio broadcasting system. Radio programming is format-oriented, attracting *station* audiences as opposed to *program* audiences. Television has been program-based until very recently. Now, however, cable television networks have followed the lead of radio in providing "format television." Students of electronic media today will be in a better position to analyze the video environment of tomorrow by understanding radio formats that began the trend toward narrowcasting from broadcasting.

The Weather Channel, CNN and Headline News, ESPN, The Nashville Network, the Home Shopping Club, VH-1, Music Television and the various pay movie services are among the cable networks which program similar content 24 hours a day. The Nostalgia Channel (and to a somewhat lesser extent Turner Network Television) concentrates on old movies. Nickelodeon programs for children. C-SPAN and C-SPAN II cablecast public affairs programming. ABC's investment into ESPN is allowing that network to take advantage of its sports division asset to expand its sports coverage (early rounds of golf tournaments are shown on ESPN with the final rounds on ABC, for example). Just as radio listeners have learned to tune into stations rather than programs, cable viewers are learning to tune into cable networks. This may eventually make the VCR recording task simpler if the viewer has a better idea of which channel is offering what programming. The VCR may cognitively represent a format as well (e.g., a quasi movie-on-demand source).

Another emergent similarity between radio and video is the ability of the audience to "zap" commercials by pushing a button. Zapping is nothing new to radio; much radio listening occurs in the car, and car radios are well suited to zapping. Interestingly, little research has been done on the radio zapping issue (Cohen, Kohl, & Greenberg, 1985, is one rare exception). Advertisers have been willing to buy radio based upon audience estimates that do not take zapping directly into account, a phenomenon that also generally describes the current video advertising environment.

A final recent development in television which is reminiscent of radio in the 1960s is miniaturization and portability. Portable television sets including color LCD receivers as small as transistor radios are becoming widely diffused. More surprisingly, in 1988 Sony introduced the "Video Walkman," a very small, portable 8mm combination VCR/3-inch color LCD television set selling for $1,000 (Lachenbruch, 1989). Should this personal video device prove successful, out-of-home VCR viewing will become a more important issue. While the television industry moves toward audience measurement via electronic meters, an increase in out-of-home portable set viewing may require increased use of radio-like measurement techniques.

Arbitron uses the same methodology to measure local radio audiences as it does in most television markets: the diary. The difference is that the radio diary is given to each *individual* whose listening behavior is being measured while the television diary is used to measure *household* viewing. The diary is also employed as a research tool in mass communication research including VCR measurement (e.g., Levy & Gunter, 1988). Understanding the strengths and weaknesses of the diary can be useful for those wishing to conduct and interpret research in the new media environment.[11]

A Note on Commercial Audience Research Sampling and Validation[12]

Commercial audience research firms often select their samples by first contacting the potential respondent by telephone whether or not the telephone is used for the actual data collection. Listed telephone households differ from unlisted homes in that they are younger, larger, more mobile, and more frequently at home. As noted by Beville (1985), these factors indicate that unlisted telephone households include heavier TV viewers. Although sampling techniques are not the subject of this chapter, it should be recognized that this is also a possible area of concern in all audience measurement.

Before discussing current audience measurement methods, a comment on the accepted method used to validate other measures is appropriate. Industry ratings methodologies are validated through what is considered the most accurate (and expensive) method available: the telephone coincidental. A telephone call is made to ask for current set tuning as well as presence of people in the room. These results are compared to those of other methods (diary, meter, telephone recall, or peoplemeter). The coincidental is limited to telephone households during nonintrusive hours of the day and still has some problem with nonresponse bias (although less of a problem than with other methods).

Radio Research: Diary Versus Telephone Recall

Unlike television, more than 40% of radio listening is out-of-home, and radio stations have budgets considerably smaller than their television brethren. These factors have contributed to the acceptance of a relatively inexpensive method of radio audience research, the personal diary that has been used since 1967 by Arbitron (Beville, 1985). The diary is intended to be filled out daily for a week by the respondent and is designed

---

[11] Although data are difficult to access, some data were made available via contacts with Arbitron and Birch. Some Nielsen data were gleaned from the trade press.

[12] Much of the material in this section is taken from Beville (1985), which is one of the only published, readily available compilations of material on commercial ratings research methodologies and validation research.

to be carried in a coat pocket. This makes recording radio listening outside the home possible.

Unfortunately, the diary has a number of limitations. Response rates for radio are as low as 30%, and can be even lower for some demographic groups (especially young men and minorities). Teens and older people are more likely to cooperate. Because the respondent is responsible for filling out listening habits, there is no control of the instrument by the research firm.[13] Cooperation with the diary falls off after the first days of the diary week. Although ideally the diary should be filled out on a daily basis, it becomes a recall activity as the diary week progresses. A diary validity question is raised when radio stations air slogans that remind the respondent to "write down" the station to which they are listening (Ross, 1988); stations that do not air such questionable reminders may be hurt. It is not inconceivable that video channels will eventually try the same tactic in local markets that use television diaries (see later).

Birch Radio Research (now Birch–Scarborough) was founded in 1978 by a former radio programmer, Tom Birch. This company has enjoyed phenomenal success in a market than saw many Arbitron competitors fail during the 1970s. Birch has perfected the telephone recall technique in which telephone respondents are asked about their radio listening habits for the previous 2 days. The strengths of this method include that recording of out-of-home listening is possible; all parts of the day are covered; an unbiased phone sample can be drawn fairly easily; it is relatively inexpensive; and unlike the diary methodology, the interviewer has control of the recall instrument. There are weaknesses: the method relies heavily on the respondent's memory; the 40% who do not respond may result in overstated listening; and inclusion of unlisted numbers is more expensive. Worth noting is that the RADAR network radio ratings service uses a *week-long,* daily telephone recall method that was praised by Beville (1985) for its high quality. Schiavone (1988) noted the differences between RADAR and custom recall research with recall exaggerating listenership.

The differences in these radio methodologies can lead to large variations between the two ratings provided for the same markets by Birch and Arbitron. The variation in resulting ratings can be seen especially in the case of album-oriented rock (AOR) stations that generally appeal to young adult men. This demographic group is also among the least

[13] A 1977 study by Pulse discovered in a test that diary keepers fell into three categories after a three day period: about a third had not made an entry into their diaries, another third had filled them out to that point, and the final third had already entered the entire week's listening with 4 days left in the period (Jablonski, personal communication, May 24, 1989).

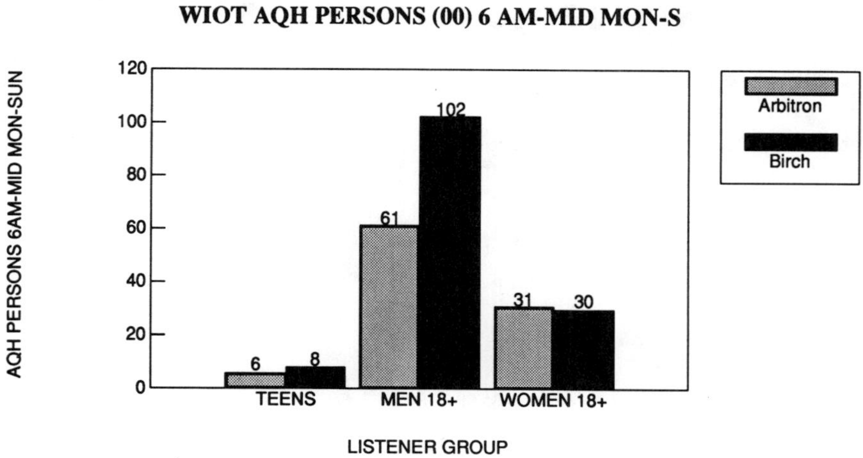

FIG. 3.1. Arbitron vs. Birch AQH persons Winter 1989. (Copyright © 1989, Birch/Scarborough Research Corp. As an Unpublished Licensed Work. All rights reserved. Printed with permission. © The Arbitron Company.)

cooperative in filling out diaries. Because young men most strongly identify with AOR stations, however, one might expect that they would fare much better via the Birch telephone recall methodology.

In Toledo, Ohio, WIOT is the only album-oriented rock station. The measurement differences are depicted graphically in Fig. 3.1 and 3.2. Figure 3.1 shows the dramatic difference between Arbitron and Birch's measures of male listeners to WIOT.[14] Figure 3.2 shows the consistency of the difference over six ratings periods (figures were not available for males only, but the differences can be expected to be at least as great). Although differences are most striking for AOR stations, other possibly systematic differences occur with other stations as well. The point here is not to determine which method is "best" (as cost considerations would also have to be taken into account), but to illustrate the potentially large impact choice of methodology can make in attempts to "simply" measure exposure to media content.

The question regarding VCR measurement is what differences will exist between telephone recall and diary methods of measuring recording and playback behavior. Is viewing of more salient rented videos more likely to be reported than viewing of less salient ones? Are there systematic differences in how the diary versus telephone recall methods are

---

[14] Average quarter hour figures are used in Fig. 3.1. This means that at any given time during the week, 10,200 men 18+ were listening to WIOT according to Birch, 6,100 according to Arbitron.

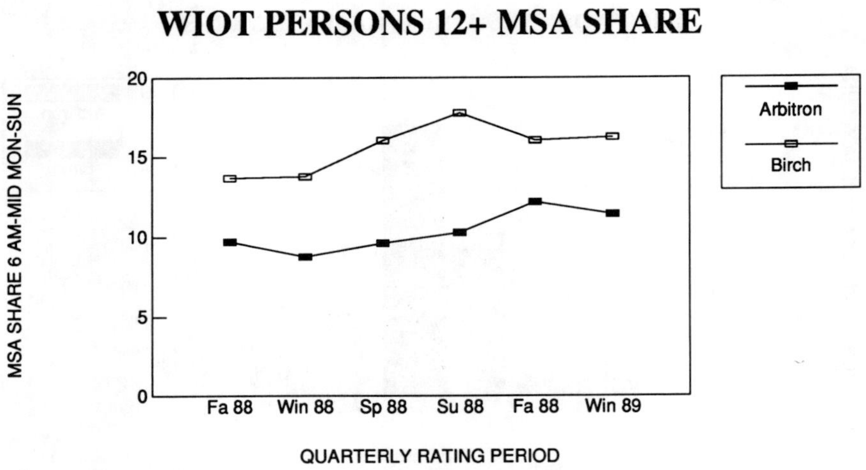

FIG. 3.2. Arbirton vs. Birch trends Toledo, OH. (Copyright © 1989, Birch/Scarborough Research Corp. As an Unpublished Licensed Work. All rights reserved. Printed with permission. © The Arbitron Company.)

susceptible to this saliency issue? These questions are worthy of further investigation especially as sales of prerecorded tapes rise (Leslie, 1989).

Television Diary

Both Arbitron and Nielsen measure local television viewing. Although television station budgets are clearly larger than those of radio stations cost is a critical factor in local television audience measurement. As discussed later, only the largest television markets can support the more expensive meter technology. The vast majority of local television markets are measured through the less costly *household* diary.

The strengths and weaknesses of the television diary are similar to those of the radio diary. Once again, the major benefit of the diary is its relatively low cost. This method appears likely to remain the primary methodology for commercial television ratings in all but the top 30 television markets for the remainder of the century. As noted by Beville (1985), the validity of the television diary also suffers from respondent's failure to report all viewing (viewing by children, teens, and young men tends to be underreported). Conditioning, in which respondents tend to modify their viewing habits as they know their behavior is being recorded, as well as a tendency to report "usual" as opposed to actual viewing are problems related to the diary (see also Kneale, 1989). There is a potential literacy problem with the diary as well with both illiterate and non-English speaking respondents. A 4-week "sweeps" period also promotes station and network hyping of programs.

The problem of *nonresponse bias* has been found to be limited (except in minority groups). A 1976 Committee on Local Television and Radio Audience Measurement (COLTRAM) study found cooperating diary households had *higher* viewing levels than non-cooperators. This was found to be somewhat offset by *response error* (the COLTRAM study found diary households *understate* actual viewing, especially that of children.) There is also the problem of entry errors in which the respondent enters an incorrect station (a problem not uncommon in radio diaries and increasingly an issue for television with its multiplicity of choices).

Out-of-Home Viewing and the Personal Diary

In response to their losses of audience to competing media including VCRs, the networks are looking for new ways to boost their ratings. One possibility is the measurement of out-of-home (O-O-H) viewing. Not only has this form of television viewing been largely ignored in the past, but the recent diffusion of small, portable television sets suggests that more viewing could now be taking place out-of-home. As of this writing, Nielsen did not include television viewing away from home; for example, college students living in dormitories and military personnel living in barracks were not included in the Nielsen sample. A Bruskin & Associates study estimated that daytime network viewing would increase 5% (with the 18–34 segment up 10%) if college students and working women were measured ("ABC pushing," 1989). There is some question whether the benefit of this additional measurement of audience is worth the expense ("Out-of-home," 1989).

Arbitron has studied the feasibility of a "personal television diary" that, like a radio diary, would attempt to measure *each* individual's television viewing in the household. Arbitron has not been persuaded that the personal diary is necessary for television. In a 1980 study, they found 4.1% of television viewing was away from home. A 1984 telephone study similarly found 96% of daily viewing at home, although 9%–10% of *weekend* viewing was not at home; men 18–34 did most away-from-home viewing. Because most portable television set viewing was also found to be done *at home*, Arbitron again believed the personal television diary was not necessary ("Out-of-home," 1989).

NBC has also been looking into out-of-home viewing. A 1980 study indicated considerable viewing of baseball playoffs and the World Series was not being measured. NBC estimated the "lost" audience to be as many as 6 million for the World Series. More recently, NBC hired an outside research firm to measure college student television viewing and concluded that they would add 33% to the measured 18–34 audience watching "Late Night with David Letterman" (Jaffe, 1988b; "NBC study," 1988).

There are two implications for VCR measurement. First, should portable VCRs become widely diffused, measurement of this use of the VCR would become necessary. Second, to the extent that VCR use (especially viewing of prerecorded tapes) is a group rather than an individual activity, out-of-home VCR measurement will also be important.

Television Meter

The television meter was introduced in the late 1950s and is used in large, local markets by Arbitron and Nielsen. Although it had been used for the Nielsen national NTI service, the older meter has been phased out and replaced by the peoplemeter. The meter has a number of inherent advantages. The meter itself requires no respondent input, registers minute-by-minute set use, provides rapid data turnaround (24–48 hours), allows further statistical analysis of viewing behavior (e.g., study of audience flow), eliminates human coding errors, gives more valid cable and independent station ratings, and employs a non-telephone based probability sample.

The greatest weakness of the meter involves its high cost. This cost weakness is compounded in that its local use leads to small local sample sizes (300–500). Other problems include relatively low cooperation rates (50% for Nielsen and less for Arbitron), possible respondent conditioning, the need for diaries to indicate individual viewing, and potential problems recording viewing of secondary sets in the household (including portable sets).

Diary Versus Meter

Although data are not readily available from Nielsen, differences between the established, diary-based ratings system and the lesser used meter system have been chronicled periodically in the trade press. Generally, network affiliated stations ratings are lower and both cable networks and especially independent stations ratings are higher on the meter. Cleveland is a case in point. Arbitron switched to a meter, whereas Nielsen continued to use only the diary. In the February 1989 ratings, Arbitron's meter showed more households using television than Nielsen's diary. The differences, which ranged from 13% to 44%, were most striking in daytime and late fringe (11 p.m. to midnight). Cable networks and independent stations were the beneficiaries, whereas network affiliates were hurt (Olevitch, 1989).

Nielsen made Minneapolis–St. Paul the 17th local television market to be metered in October 1988, and independent station KMSP-TV benefited from substantial audience share increases in the following dayparts ("Meters rattle," 1988):

- 3–5 p.m.: share up 67% from 12 to 20% (moved from #4 to #2 station)
- 5–6:30 p.m.: 23% share (up 77%)
- 7–9 p.m.: from 5 to 13%
- 10:30–11:30 p.m.: from 5 to 11%

Arbitron research shows the following differences between the local meter and the diary (Jaffe, 1988a). There is a "salience" factor: Fewer important or popular shows are recorded in the diary. The diary misses "short tuning" (partial quarter hour viewing). The diary does not pick up "tuning without viewing." "Diary fatigue," in which diary keeping is less reliable, sets in during the latter part of the diary week. A study by Statistical Innovations found similar results. Diary underreporting is higher in daytime than early fringe (late afternoon/evening). Films are underreported, sports and news less so. The least underreporting occurs on Thursday, the first day of the Nielsen diary; declines increase to a maximum on the following Wednesday.

Independent television stations and cable television networks get higher ratings on peoplemeter based services than they do via diary measurements (Jaffe, 1988a, 1988b). In 13 metered markets in November 1986, independent stations had ratings that were 29% higher than in diaries, and network affiliates' viewership was 10% lower. A study by the Independent Television Association found seven significant variables when comparing meters versus diaries: household size, station type (affiliate or network), VCR presence, program type (movies underreported), day of week (weekends underreported), daypart (daytime and early fringe underreported), and time within daypart ("New ratings scale," 1988). The independent television industry wants to *calibrate* diary ratings to the meter results.

Given a choice, advertising agencies have generally migrated toward meter-based services where they are available. The direct implication for VCR measurement is important as more advertisers experiment with placing commercial announcements on prerecorded cassettes. Once again, one would expect that diary-based measures of exposure to those messages will differ from electronic-based measures of that exposure.

Peoplemeters

The television industry has experienced a revolution in audience measurement in the late 1980s (Kaplan, 1988; Kneale, 1988b; McKenna, 1988; Sylvester, 1988). By the mid-1980s, it was becoming clear that the old meter and diary-based system used by Nielsen for decades to estimate national audiences was about to be replaced (Advertising Research Foundation, 1987; Beville, 1986c). Advertisers in particular were concerned with the diary and its limited sample in the new video environment (Stoddard, 1988). While Nielsen began testing a prototype peoplemeter in 1978, Britain's AGB had been using its people system in Europe. AGB announced its intent in 1983 to enter U.S. TV research with its peoplemeter that was already in use in Europe. Nielsen responded by testing its national peoplemeter in 1984. After successfully testing its peoplemeter in Boston in 1985, AGB announced it would begin a national service in the Fall of 1987. Not coincidentally, Nielsen followed AGB's lead with an identical announcement, and the era of the peoplemeter began in September 1987 (Honomichl, 1987).

Broadcaster objections that the peoplemeter was not ready for the marketplace seemed justified as differences between the AGB and Nielsen services existed, but faded as samples grew. One of the few differences that existed between the two services was AGB's attempt to start reporting VCR use. AGB produced a report in March 1988 which showed both recording *and* playback use of VCRs while Nielsen continued to ascribe VCR recording as viewing. By the Fall of 1988, support for AGB waned, and the company ceased operations following the withdrawal of CBS's backing. Once Nielsen caught up with AGB in rolling out its peoplemeter service, advertising agencies saw no reason to switch from Nielsen to AGB. Because the agencies stayed with Nielsen, so did the broadcast networks.[15]

Arbitron tested its own version of a peoplemeter in Denver and announced in 1989 that the system would be expanded by the end of 1989. Three local markets (Minneapolis, St. Louis, and Sacramento) were to be added to Denver. ScanAmerica represents what is expected to be the next wave in commercial market research: "single-source data," the simultaneous collection of both television viewing and product purchase

[15] Advertising agencies can be the critical link to audience research service viability. Birch Radio marketed its services to agencies to create the needed demand for eventual adoption by radio stations (Jablonski, personal communication, May 24, 1989). Advertising agencies in Cleveland dropped the local, diary-based Nielsen service in favor of the meter-based Arbitron service, and television stations had no choice but to follow the advertisers' lead (Olevitch, 1989).

behaviors (Gold, 1988). These data will be collected from the same panel of respondents (Schwartz, 1989; Zackon, 1987). Arbitron pays respondents about $1 per day, and they are expected to record all product purchases via the UPC codes on packages that are read by a light pen. Nielsen, too, is working on its own single-source system which is expected to be different from its ratings system (Gold, 1988; Jaffe, 1989).

Peoplemeter Problems

The peoplemeter is not without its problems. NBC research vice president Bill Rubens has criticized the fatigue factor involved in expecting household members to press buttons ("NBC exec believes," 1988; Rubens, 1984). He also expressed a concern with Nielsen's panel that he wanted to see turnover every 2 years rather than the 5 years originally proposed by Nielsen (Mandese, 1987).

CBS vice president David Poltrack has criticized the Nielsen move to peoplemeters from the outset (Bierbaum, 1988a; Poltrack, 1988). CBS traditionally benefited from the diary technique as its programs tended to skew toward older and more rural television households than the general audience; if these people were more likely to fill out diaries, then CBS's ratings could have been overstated historically. Poltrack pointed out a key difference in the meter versus the diary by examining the battle of CBS's "Dallas" versus NBC's "Miami Vice" in November 1987. Poltrack found that "Dallas" had more viewers than "Miami Vice," but that the percentage differences indicated great variance between methods:

- 33% in Nielsen diaries
- 30% in Nielsen old meter system
- 29% in Arbitron diaries
- 26% in telephone coincidental
- 5% in Nielsen peoplemeter

Poltrack (1987) explained that the differences were due to differences in both methodology and sampling.

A major difference exists between the peoplemeter sample and a diary sample. Although respondents are only asked to keep a diary for 1 week's time, the peoplemeter panel stays in place. This difference, of course, also exists with local household meter samples as well. Soong (1988) concluded that the peoplemeter "is considerably more reliable than a diary sample with the same number of households" (p. 55). Although this bodes well for eventual VCR measurement, subsequent differences in the measurement techniques must be reconciled.

The Peoplemeter, Children, and Videos

A large controversy with the peoplemeter developed over its ability to measure children's viewing of television (Friedman, 1988; "Kids' viewing down," 1988; Mandese, 1987; Rovner, 1989; Teinowitz, 1988; "Understanding children's," 1989). Simply stated, although children are in the room watching television, they do not seem to push the appropriate buttons. The result has been that Saturday morning program ratings were down 25%–30% in the Fall of 1987. Clearly the implications for commercially sponsored children's programming are great.

MTV Networks has sponsored two Nielsen studies of its audience that have indicated that actual viewership by children, teens, and young adults is higher than figures generated by the peoplemeter. The latest telephone coincidental involving 75,000 viewers was conducted in the Spring of 1989 and indicated that the audiences for each of these groups was higher for Saturday morning shows, "Nickelodeon," and weekday syndicated children's shows (Walley, 1989b). The coincidental found audience increases that varied from 11% to 34% depending on youth segment and daypart. Unfortunately for producers of children's programming, advertisers are generally unwilling to make adjustments in the peoplemeter data ("Understanding children's viewing," 1989).

The implications for measuring children's use of the VCR are equally important. One would expect the peoplemeter to similarly underestimate viewing of prerecorded cassettes by young people. This could have an economic and social impact. Economically, advertisers may be less willing to subsidize the cost of children's videos if their use is not measured accurately. Similarly, researchers will have a difficult time discerning what tapes are being watched by children.

R. D. Percy & Company Infrared Passive Peoplemeter

Perhaps the most technologically sophisticated peoplemeter rating system was that developed by Roger Percy's Seattle-based research firm. The system was tested extensively in New York and was developed to measure commercial ratings. The system used an infrared scanner called the Voxbox 1200 to record who was in the room at any given time. The Percy passive peoplemeter was called the Voxbox 1200 and it recorded both the channel being watched as well as both channel and volume changes. The heat-sensing peoplemeter also detected when someone entered or left the room (Kneale, 1988b). Importantly, the meter could detect when a VCR recorded a program, when the program was played back, and whether or not commercials were being deleted or fast-forwarded.

The system measured television viewing *every second* in about 1,000 homes in 29 counties in the New York metropolitan area in 1988. This frequent measurement was intended to measure commercial viewing for the first time. Although the system was supported in test phases by NBC, CBS, and advertisers including Coca-Cola and Kraft, many questioned the system's ability to accurately detect room movements during 30 and, especially 15-second commercials (Blau, 1988; Kneale, 1988c). Percy intended to provide the service in the largest local television markets (Buckman, 1987). Whether due to being ahead of its time or (as others have commented) simply not up to the technical task, Percy failed to survive in the audience measurement arena. Thus, another opportunity to measure VCR use was lost.

The Passive Peoplemeter and the Future

All of the existing measurement techniques reviewed involve some degree of direct cooperation from members of the sample household. In a truly ideal world, all that would be needed is the permission of the members of the household to be included in the sample with no further effort required on their part. Apparently Percy's technology was not the answer. Collet (1987) experimented with videotaping the in-house audience, but the problems with privacy and research validity are clear. Some have suggested that household members wear some kind of "electronic tag," but even this would require a certain amount of cooperation on the person's part. The technical challenges of the passive peoplemeter remain great (Lu & Kiewit, 1987).

Nielsen announced in 1989 that a passive peoplemeter was under development. According to the company, the goal of the system is to eliminate active involvement of people in the sample households. If successful, it will increase ratings accuracy while lessening cooperation fatigue. The system is under development at the David Sarnoff Research Center and depends on digital image tracking and recognition technology. The system would detect objects in its visual field, compare them with those stored in memory, and (hopefully) correctly identify the persons in the room. Nielsen itself does not expect this technology to be in use before 1993 ("Nielsen to develop," 1989). The technology involved in the peoplemeter would allow it to measure viewing of commercials as well. Nielsen may also use the passive system to develop its own single source of data including product purchase behavior (Walley, 1989a).

Arbitron was also planning to work with the French company Telemetric on a passive peoplemeter system. The device is expected to be tested in the United States by Arbitron and involves a form of optical technology that detects movement. The system is called Motivac and was already in use in Paris in 1989. Further technical development is necessary as it can only identify individual bodies but not their identities ("Arbitron close," 1989). The system also records whether or not a VCR is in use (Murrow, 1989). The passive peoplemeter may also offer the most hope for the critical task of measuring children's video use.

It should be noted, however, that some other new sensing technology may be developed even before a pattern recognition system is readied. It would appear that concerns about privacy by people in the sample households will be at least as much of a barrier as any obstacles surrounding the technology. Perhaps a study of viewers' perceptions of the privacy implications of an existing technology, pay-per-view billing of socially unacceptable material (whether pornographic material or even a wrestling event) would be a good precursor to understanding potential resistance to the passive peoplemeter. There may still be more perceived anonymity on a bill than in the knowledge that one's image is being recognized by a machine in the living room.

CONCLUSIONS

Commercial research firms have the capacity to objectively measure and track VCR usage data in a more valid manner than past methodologies have allowed. This chapter has highlighted differences in VCR use depending on the type of measurement used, and shown that there are contradictions. Perhaps private research firms will break with tradition by making data available for analysis by academics with no particular point of view to protect. Mass communication scholars have often had to rely on inexpensive, survey research techniques that may include comparatively invalid self-reports of video usage behavior. Perhaps the continuing progress of technology may one day bring the cost of measuring the audience down to accessible levels. Unless and until that day comes, one hopes the commercial research firms are archiving their data on VCR use. Scholars may have to be satisfied with post hoc analyses of previously proprietary data. In the academic pursuit of knowledge, however, the analysis of the most valid data available should be sought.

This chapter represents an effort to help point VCR researchers in that direction.

ACKNOWLEDGMENT

Portions of the research presented here were supported by grants from the National Association of Broadcasters and the Kaltenborn Foundation.

REFERENCES

ABC pushing for out-of-home viewing measure on broad basis. (1989, July 24). *Television/Radio Age,* p. 11.

A. C. Nielsen Company. (1989). *1989 report on television.* Northbrook, IL: Author.

Arbitron close to deal on passive people meter. (1989, June 12). *Television/Radio Age,* p. 18.

Advertising Research Foundation. (1987). *Peoplemeters: Evolving a constructive industry course.* New York: Author.

Benson, J. (1988, November 28). Trouble corralling the grazers. *Advertising Age,* p. S-4.

Beville, H. M. (1985). *Audience ratings: Radio, television, and cable.* Hillsdale, NJ: Lawrence Erlbaum Associates.

Beville, M. (1986a, February 17). VCR usage patterns begin to emerge. *Electronic Media,* p. 26.

Beville, M. (1986b, March 17). Taking a look at the impact of VCRs. *Electronic Media,* p. 66.

Beville, M. (1986c, November 10). Industry is only dimly aware of people meter differences. *Television/Radio Age,* pp. 78–81, 110.

Bierbaum, T. (1988a, June 17). Peoplemeters are biased against CBS, sez Poltrack. *Variety,* pp. 65, 85.

Bierbaum, T. (1988b, August 24). A. C. Nielsen set to measure video playback patterns. *Variety,* pp. 1, 115.

Blau, E. (1988, April 25). People-metering for commercials. *The New York Times.*

Buck, S. (1988). Television audience measurement research—yesterday, today and tomorrow. *Journal of the Market Research Society, 29*(3), 265–278.

Buckman, A. (1987, March 30). Peoplemeters: Industry braces for impact of new technology. *Electronic Media,* pp. T1, T11.

Cassata, M., & Irwin, B. (1989, April). *Commercial avoidance behavior of VCR users during time-shifted programs: The effects of zipping on the recognition and recall of advertisements.* Paper presented at the meeting of the Broadcast Education Association, Las Vegas, NV.

Cohen, E., Kohl, L., & Greenberg, B. (1985, October). The growing threat of dial zapping. *Radio Only,* p. 20.

Collet, P. (1987, Fall). The viewers viewed. *Et cetera,* pp. 245–251.

Durand, J. (1988). Research without frontiers: Towards Europe-wide television audience measurement. *EBU Review, 39*(3), 11–16.

Ehrenberg, A. S. C., & Wakshlag, J. (1987). Repeat-viewing with people meters. *Journal of Advertising Research, 27*(1), 9–13.

Friedman, W. (1988, March 28). Metering the TV wars: who's ahead; why? *Cablevision,* pp. 37–40, 44.

Gay, V. (1987, April 6). Nets win battle for VCR playback ratings. *Advertising Age,* p. 8

Gelman, M. (1987, May 20). VCRs shaping prime time ploys. *Variety,* pp. 49, 80.

Gold, L. N. (1988). The evolution of television-sales measurement: Past, present and future. *Journal of Advertising Research, 28*(3), 19–24.

Goodhardt, G. J., Ehrenberg, A. S. C., & Collins, M. A. (1987). *The television audience: Patterns of viewing (An update).* Brookfield, VT: Gower.

Hadlock, W. G. (1988, November 28). Another bad year for TV's goliath's. *Advertising Age,* pp. S-4, S-6.

Henke, L. L., & Donahue, T. R. (1989). Functional displacement of traditional TV viewing by VCR owners. *Journal of Advertising Research 29*(2), 18–23.

Hey, K. R. (1987, October). We are experiencing network difficulties. *American Demographics,* pp. 38–40, 69–70.

Hickey, N. (1988, March 19). The verdict on VCRs (so far): They're changing what you see. *TV Guide,* pp. 12–14.

Home tape study cues webs grins. (1985, December 11). *Variety,* pp. 1, 143.

Honomichl, J. (1987, July 27). Collision course: Stakes high in people meter war. *Advertising Age,* pp. 1, 68, 70.

Huff, R. (1988, December). People meters and cable. *Marketing & Media Decisions,* pp. 38–40.

Hurwitz, D. (1984). Broadcast ratings: The missing dimension. *Critical Studies in Mass Communication, 1,* 206–215.

Jaffe, A. J. (1988a, June 27). Outlook not good on closing the meter-diary gap. *Marketing & Media Decisions,* p. 35.

Jaffe, A. J. (1988b, October 31). TV diary project underway. *Television/Radio Age,* pp. 43–44.

Jaffe, A. J. (1989, March 20). Can Arbitron make it? *Television/Radio Age,* pp. 30–32.

Kaplan, B. (1988, March 7). *The effect of people meter measurement on the advertising marketplace.* Speech presented to the annual conference of the Advertising Research Foundation, New York.

Kaplan, B. M. (1985). Zapping—The real issue is communication. *Journal of Advertising Research, 25*(2), 9–12.

Kavanaugh, M. (1989, February 9). Thinly spread viewers baffle head-counters. *Marketing,* p. 11.

Kidder, Peabody, & Co. (1988, September 6). *Television programming: People watch television programs, not television stations.* Industry report.

Kid's viewing down according to study. (1988, January 13). *Variety,* p. 39.

Klopfenstein, B.C. (1988, May). *The emerging VCR household: Relationships among ownership, demographics, and usage patterns.* Paper presented to the International Communication Association, New Orleans, LA.

Klopfenstein, B.C. (1989a). The diffusion of the VCR in the United States market. In M. R. Levy (Ed.), *The VCR age: Home video and mass communication* (pp. 21–39). Beverly Hills, CA: Sage.

Klopfenstein, B. C. (1989b, November). *Looking toward future VCR use: An examination of VCR use in four household adopting groups.* Paper presented to the Speech Communication Association, San Francisco, CA.

Klopfenstein, B.C., & Swanson, D.A. (1987, May). *An analysis of VCR adopter characteristics and behavior.* Paper presented to the International Communication Association, Montreal.

Kneale, D. (1988a, April 18). As TV season ends, ratings show networks lost millions of viewers. *The Wall Street Journal,* p. 20.

Kneale, D. (1988b, April 25). Using high tech tools to measure audience. *The Wall Street Journal,* p. 21.

Kneale, D. (1988c, April 25). Zapping of TV ads appears pervasive. *The Wall Street Journal,* p. 21.

Kneale, D. (1989, March 3). How to shun friends and influence shows: Notes from a week as a Nielsen household. *The Wall Street Journal,* p. B1.

Knight, B. (1989, March 8). Surprise! NBC wins the sweeps. *Variety,* pp. 43, 54.

Knobeisdorff, K. E. (1988, January 6). For TV networks, 'people meter' is a profit eater. *Christian Science Monitor,* pp. 10–11.

Krugman, D. M. (1985). Evaluating the audiences of the new media. *Journal of Advertising, 14*(4), 21–27.

Lachenbruch, D. (1989, January). Will the Video Walkman give Sony the last laugh in the format wars? *Consumer Electronics,* p. 74.

Lerner, D. (1988, June). Understanding erosion: Part I. *Marketing & Media Decisions,* pp. 112–116.

Leslie, C. (1989, January 11). Doomsayers of yore, note: People are buying cassettes. *Variety,* p. 39.

Levy, M. (1980a). Program playback preferences in VCR households. *Journal of Broadcasting 24*(3), 327–336.

Levy, M. (1980b). Home video recorders: A user survey. *Journal of Communication, 30*(4), 23–27.

Levy, M. (1981). Home video recorders and time shifting. *Journalism Quarterly, 58,* 401–405.

Levy, M. (1983). Time-shifting use of home video recorders. *Journal of Broadcasting, 27*(3), 263–268.

Levy, M. R. (1987). Some problems of VCR research. *American Behavioral Scientist, 30*(5), 461–470.

Levy, M. (1988, April). *Why VCRs aren't pop-up toasters: Theoretical issues in VCR research.* Paper presented to the annual meeting of the Broadcast Education Association, Las Vegas, NV.

Levy, M., & Gunter, B. (1988). *Home video and the changing nature of the television audience.* London: John Libbey.

Lewin, K. (1988, December). Getting around commercial avoidance. *Marketing & Media Decisions,* pp. 116–122.

Lieberman, D. (1987, December 7). How do you spell pain? ABC, NBC, and CBS. *Business Week,* pp. 128, 130.

Lindstrom, P. B. (1989). Home video: The consumer impact. In M. R. Levy, (Ed.). *The VCR age: Home video and mass communication* (pp. 40–49). Beverly Hills, CA: Sage.

Lippman, J. (1989, January 11). Webs say good riddance to the year of disappointment. *Variety,* pp. 5, 9.

Lu, D., & Kiewit, D. (1987). Passive people meters: A first step. *Journal of Advertising Research, 27*(3), 9–14.

Luchter, L. (1988, March 21). AGB puts VCR use at 7 hours weekly. *Multichannel News,* pp. 1, 49.

Luyken, G. M. (1987). The VCR explosion and its impact on television broadcasting in Europe. *Columbia Journal of World Business, 22,* 65–70.

Mahler, R. (1987, August 17). VCRs vs. TV stations. *Electronic Media,* pp. 1, 30.

Mandese, J. (1987, September). The all new ratings game. *Marketing & Media Decisions,* pp. 39–45.

May penetration. (1989, July 3). *Broadcasting,* p. 45.

McKenna, W. (1988). The future of electronic measurement technology in U.S. media research. *Journal of Advertising Research, 28*(5), RC-3–RC-7.

McSherry, J. (1988, December). The impact of multi-set viewing on media planning. *Marketing & Media Decisions,* p. 132.

Meehan, E. (1984). Ratings and the institutional approach. *Critical Studies in Mass Communication, 1,* 216–225.

Meters rattle Minneapolis-St. Paul. (1988, November 14). *Television/Radio Age,* p. 18.

Metzger, G. (1986). CONTAM's VCR research. *Journal of Advertising Research, 26*(2), RC-8–RC-12.

Murrow, D. (1989, June 5). French test passive people meter. *Advertising Age,* p. 76.

Nayeri, F. (1988, December 1988). People-meter makes French debut. *Variety,* p. 39.

NBC exec believes people-meter research is at times invalid. (1988, January 13). *Variety,* pp. 1, 48.

NBC study: Latenight viewing 33% more than first reported. (1988, February 24). *Variety,* p. 480.

New ratings scale for diary services sets INTV debut. (1988, January 6). *Variety,* pp. 33, 88.

Nielsen Media Research. (1988). Videocassette recorders in the 1980s. *Nielsen Newscast,* pp. 2–9.

Nielsen to develop passive people meter. (1989, June 5). *Broadcasting,* p. 31.

Nielsen people meter service launched. (1987). *Nielsen Newscast, 3,* 2–6.

Olevitch, L. (1989, June). *Discrepancies between diary and meter ratings data in the Cleveland market.* Presentation made to audience research seminar, Bowling Green State University, Bowling Green, OH.

Ostroff, J. (1989, February). Television ratings and the theory of evolution. *Marketing & Media Decisions,* pp. 84–86.

Out-of-home measurement issue heats up. (1989, June 26). *Television/Radio Age,* p. 12.

Paskowski, M. (1988, March 21). AGB studies VCR use. *Electronic Media,* p. 20.

Philport, J. (1989, April 13). Address given at Bowling Green State University, Bowling Green, OH.

Poltrack, D. F. (1987). The people meter: Its time has come—too soon. In *People meters: Evolving a constructive industry course.* New York: Advertising Research Foundation.

Poltrack, D. F. (1988). Living with people meters. *Journal of Advertising Research, 28*(5), RC-8–RC-10.

Reed, D. (1989, January 5). Satellite proves tough for audience watchers. *Marketing,* p. 11.

Rosenthal, E. M. (1987, May 25). VCRs having more impact on network viewing, impact. *Television/Radio Age,* pp. 37–39, 68–70.

Rosenthal, E. M. (1988, September 19). Cable operators stage a comeback. *Television/Radio Age,* pp. 46–48.

Ross, S. (1988, December 10). Stations fill air waves with diary-aware imagery. *Billboard,* pp. 10, 15, 17.

Rovner, R. (1989). *The ability of the Nielsen people meter to accurately measure children's television viewing.* Unpublished senior thesis, College of Communication, Boston University, Boston, MA.

Rubens, W. S. (1984). High -tech audience measurement for new-tech audiences. *Critical Studies in Mass Communication, 1,* 195–205.

Schiavone, N. P. (1988). Lessons from the radio research experience for all electronic media. *Journal of Advertising Research, 28*(5), RC-11–RC-15.

Schwartz, J. (1989, January). Back to the source. *American Demographics,* pp. 22–26.

Sims, J. B. (1989). VCR viewing patterns: An electronic and passive investigation. *Journal of Advertising Research, 29*(2), 11–17.

Smith, D. L. (1989, March 6). The zip, zap and graze craze. *Advertising Age,* p. 20.

Soong, R. (1988). The statistical reliability of people meter ratings. *Journal of Advertising Research, 28*(1), 50–56.

Sternberg, S. (1989, January). VCRs: A new medium, a new message. *Marketing and Media Decisions,* pp. 81–84.

Stitt, R. (1989, January 9). Reports on broadcast television's death greatly exaggerated. *Television/Radio Age,* p. 63.

Stoddard, L. R. (1988). The history of people meters: How we got to where we are (and why). *Journal of Advertising Research, 28*(5), pp. RC-10–RC-12.

Study finds little TV grazing. (1989, July 31). *Advertising Age,* p. 35.

Survey: Ad dollars going to cable, indies, shift began mid-'80s. (1988, August 24). *Variety,* p. 92.

Swanson, D. A., & Klopfenstein, B. C. (1987, December). How to forecast VCR penetration. *American Demographics,* pp. 44–45.

Sylvester, A. (1988, September). Television-audience measurement in transition. *Marketing & Media Decisions,* pp. 84, 88.

Teinowitz, I. (1988, July 18). People meters miss kids. *Advertising Age,* p. 4.

Teinowitz, I. (1989, February 27). Nielsen to track videos. *Advertising Age,* p. 28.

Trachtenberg, J. A. (1988, September 19). Diary of a failure. *Forbes,* pp. 168, 170.
Understanding children's viewing on TV: Agreed, but what to do? (1989, June 26). *Television/Radio Age,* p. 14.
VCRs in 65% of homes. (1989, June 26). *Television Digest, 29,* p. 17.
VCRs play heavy role in viewing. (1988, October 31). *Television/Radio Age,* p. 39.
Von Hippel, E. (1986). Lead users: A source of novel product concepts. *Management Science, 32*(7), 791–805.
Walley, W. (1988, October 17). Viewer declines force TV news to boost ads. *Advertising Age.* p. 32.
Walley, W. (1989a, June 5). Nielsen's new meter may give ratings for ads. *Advertising Age,* pp. 1, 76.
Walley, W. (1989b, June 26). MTV Networks versus the people meter. *Advertising Age,* p. 59.
Webster, J. G. (1989a). Assessing exposure to the new media. In J. L. Salvaggio & J. Bryant (Eds.), *Media use in the information age* (pp. 3–19). Hillsdale, NJ: Lawrence Erlbaum Associates.
Webster, J. G. (1989b). Television audience behavior: Patterns of exposure in the new media environment. In J. L. Salvaggio & J. Bryant (Eds.), *Media use in the information age* (pp. 197–216). Hillsdale, NJ: Lawrence Erlbaum Associates.
Webster, J. G., & Wakshlag, J. (1985). Measuring exposure to television. In D. Zillmann & J. Bryant (Eds.), *Selective exposure to communication* (pp. 35–62). Hillsdale, NJ: Lawrence Erlbaum Associates.
Weekly TV viewing down last season. (1988, January 6). *Variety,* p. 80.
Williams, F., Rice, R. E., & Rogers, E. M. (1988). *Research methods and the new media.* New York: The Free Press.
Yorke, D. A., & Kitchen, P. J. (1985). Channels flickers and video speeders. *Journal of Advertising Research, 25*(2), 21–25.
Zackon, R. (1987). Commercial audience measurement and single source data: What the 1989–1990 TV season may look like. In *People Meters: Evolving a constructive industry course.* New York: Advertising Research Foundation.
Zahradnik, R. (1986, June). Rewinding VCR penetration. *Marketing & Media Decisions,* p. 28.
Zoglin, R. (1988, October 17). The big boys' blues. *Time,* pp. 56–61.

II

The Relationship of VCRs to Theoretical Frameworks: Testing, Extending, or Maintaining Existing Media Theories

4

Audience Activity and VCR Use

Carolyn A. Lin
Southern Illinois University

Videocassette recorders (VCRs) were first brought to the consumer market by Sony Corporation in 1975 (*Advertising Age,* 1975). Along with cable TV, VCRs represent a home entertainment medium that provides TV viewers with expanded viewing options and increased technical controls over their viewing process. The development of cable TV enabled TV viewers to become a more selective audience by offering more diversified programming. It was, however, the introduction of VCRs that emancipated the TV audience from being a *passive viewer* to an *active user.* Through the use of the VCR's technical features, TV viewers were able to actively select their viewing options from sources beyond their regular TV program schedule, in addition to timing their viewing decisions.

As VCR penetration surpasses 60% of American TV homes (the most rapid growth rate for any electronic entertainment medium), this phenomenon is not unique. The quest for VCR ownership seems to be a worldwide phenomenon, especially among nations with much less diversified media environments, where VCR penetrations are typically higher than the United States (Lin, 1987). VCR use thus helps supplement rather than supplant what the TV programming schedule lacks, in terms of content and time flexibility.

Early research suggested that VCR users are a rather *active* audience that intend to utilize the VCR to its full potential as a home entertainment device. For instance, Levy (1983) reported that the average VCR household made about four recordings, played back between three and four tapes, and watched less than one tape, which had been bought, rented, or borrowed (per week). Another study (Levy & Fink, 1984) revealed that VCR users time-shift the broadcast schedule for more convenient viewing; these time-shifting activities were perhaps indicative of VCR users' intention to either maintain or increase their media-use gratification. However, one study reported that approximately 20% of cable subscribers who owned a VCR did not have their VCR hooked up to cable (or did not know if they had it hooked up) and another 30% had trouble hooking up their VCR (*Cablevision,* 1985). As a promotion scheme, a multiple system operator (Group W Cable) launched a campaign to help its subscribers hook up their VCRs to promote the compatibilities between cable and VCR technologies (*Advertising Age,* 1985).

Although user profile studies are able to fulfill descriptive purposes of VCR use activities, they may not be useful in gauging the psychological and behavioral aspects of such activities. Some recent studies have explored the theoretical implications of VCR use in relation to media uses and gratifications (e.g., Rubin & Bantz, 1987) as well as audience activity (Levy, 1987; Rubin & Bantz, 1987).

Past research has extensively documented the theoretical applications of uses-and-gratifications perspectives. However, the construct of *audience activity,* which focuses on the impact of media-use processes on media-use motivations and future media-use behavior, still needs further examination. This chapter attempts to explore various VCR-use activities and their relations to the overall audience activity paradigm.

AUDIENCE ACTIVITY PARADIGM

Amidst the many uses-and-gratifications perspectives studied during the 1970s, Blumler (1979) expanded the notion of an "active audience" within the context of "audience activity" as an intervening factor in the media-use process. He articulated the potential audience activities that may occur before exposure, such as *selectivity* (e.g., knowing the viewing options or making viewing plans in advance) and *intentionality* (or expectations for media exposure experiences). *Viewing attention* is a cognitive activity that could affect the media-use process during exposure; and *utility* (i.e., utilizing media materials for interpersonal communication purposes or absorbing them into other activities such as purchasing behavior) reflects behavioral involvement after exposure.

Several researchers endorsed the *active audience* concept; they maintained that the "active-ness" that occurred during the media-use process could explain subsequent media-use behavior (Galloway & Meek, 1981; Palmgreen & Rayburn, 1979). Wenner (1985) considered that audience activities can also be the key to explaining the transactional links between gratifications sought and gratifications obtained.

Opponents of the active audience concept proposed the view of an *obsti-*

nate audience, which assumes external factors such as mealtime or work schedule instead of internal factors such as needs or motivations dictate media-use behavior (Bauer, 1964; Bogart, 1965). However, proponents of an active audience pointed out that audience members often had strong and loyal preferences among equally available mass media, indicating that external factors (or habits) alone could not determine media use activity. For instance, McGuire (1974) suggested that although external factors dominate the initial stage of media exposure, subsequent exposure is motivated by internal forces. Blumler (1979) advanced a similar assertion in which both external circumstances and internal motivations are the forces motivating different stages of the media-use process.

Based on these theoretical premises, a number of researchers examined the audience activity construct. The common theme shared by these studies indicates that audience activities comprise the *selectivity, involvement,* and *utility* aspects of the media-use process, which may reflect an audience member's cognitive, affective, and behavioral involvement with the process itself. Activities such as knowing or planning one's viewing decision ahead of time (Lemish, 1985; Levy & Windahl, 1985) reflect an extensive level of cognitive involvement with media use as a selective process. Other activities, such as thinking about media content during or after viewing (Greenwald & Leavitt, 1984) and attention paid during viewing (Kellerman, 1985), are all relevant to certain cognitive processes that can depict a viewer's involvement with media content.

By contrast, activities such as parasocial interaction (Levy, 1979; Rubin & McHugh, 1987; Rubin & Perse, 1987) with media characters or content being viewed, are indicative of a certain kind of affective involvement between viewers and the media content under consumption. Moreover, activities such as interpersonal discussion of media content during or after exposure (Lemish, 1985; Levy & Windahl, 1985; Standford, 1984) are also examples of behavioral reactions toward the media content of the utility dimension.

Other aspects of the utility dimension postulated by Blumler (1979), such as purchasing activities motivated by media advertising or other activities encouraged by media messages, are operational examples of the audience's behavioral involvement with the media content consumed. However, to date no research effort has been devoted to exploit this theoretical assumption.

VCR-USE ACTIVITY

Uses of VCRs reported by audience members primarily include time-shifting, pre-recorded tape viewing, and video library building (Donohue & Henke, 1988; Harvey & Rothe, 1986; Levy & Fink, 1984). Although

VCRs are also used to replay home movies, such a practice should not be prevalent until the distribution of videocameras dramatically rises (see Vale, chapter 11, this volume).

Recordings made for time-shifting purposes are for "delayed viewing" purposes, or viewing at a later more convenient time. Time-shifting usually means viewing planning in advance, if program recordings are to be made on time. During the recording period, the audience member can either watch another channel not being recorded or not watch TV at all. According to Rubin and Bantz (1987), time-shifting activity may enhance interpersonal communication, when the purpose of timing viewing schedules is to enable more audience members to watch the recording together.

Building a video library also requires advance planning. Unlike the *delayed viewing* purpose of time-shifting recording, video library copies are made for *delayed playback* purposes, which can mean a long interval between recording and playback time. Video library recordings can be produced while the audience member is watching the program being recorded, watching a program not being recorded, or not watching TV at the time of recording. Once again, as several authors (Harvey & Rothe, 1986; Rubin & Bantz, 1987) pointed out, socialization or interpersonal communication may be related to the use of video library copies when families or friends get together to watch videos in a social setting.

Pre-recorded video rentals provide alternative viewing options that are unavailable on the current TV program schedule. These rentals can include movie titles, specialized entertainment (e.g., concerts), or other special purpose videos (e.g., do-it-yourself, educational or instructional titles). The viewing of these pre-recorded programs may preempt the viewing of regularly scheduled programs. As the recent trend indicates, more VCR users rent pre-recorded tapes than record programs for time-shifting purposes (A. C. Nielsen, 1988). Such a tendency might have been a result of the multiplicative increase of available movie titles on video.

Regardless of the VCR use purposes, recording and replaying tapes are the primary VCR-use activities. An important dimension observed from such activity is the behavior of commercial pausing during recording or commercial zipping during playbacks (Yorke & Kitchen, 1985). Commercial pausing, which requires constant manipulations of a remote-control device, pauses the VCR to avoid recording commercials. Commercial zipping, on the other hand, can be done by merely fast-forwarding the VCR to skip the viewing of the commercials recorded. Both activities parallel commercial zapping behavior during TV viewing, which is accomplished by switching to watch another channel during commercials—an activity reportedly practiced by more than half of TV viewers (Heeter & Greenberg, 1985).

THEORETICAL ASSUMPTIONS

It is thus logical to assume that recording and replaying activities, as well as commercial pausing and zipping, are part of the overall audience activity paradigm. These activities are perhaps either dependent on or related to a range of other audience activities, such as selectivity, involvement, and utility. Generally, these audience activities are performed for the purposes of maximizing one's TV viewing enjoyment, where TV viewing reflects the use of television as a medium to show different programs.

From an audience activity perspective, VCR-use activities are the *process* variables that enable the completion of a VCR-use cycle and provide the VCR-use or TV-viewing experience that may influence subsequent media consumption behavior. As suggested by the few extant empirical studies that examined VCR use and audience activity, prompted by their technological endowment, VCR users are identified as actively engaged in various types of audience activities. For instance, based on an Israeli sample, Levy (1987) found that the majority of audience members tended to be selective in terms of program recording choice during the pre-exposure period, somewhat involved with the video content during the exposure period, and capable of deriving interpersonal communication utility from the VCR-use experience during the post-exposure period. Greenberg and Lin (1989) found that when the degree of involvement with audience activities was compared among a group of adolescent viewers, VCR users were clearly a more active audience than non-VCR users.

METHODS

Data for this study were collected during the spring of 1987. A telephone survey was conducted in a Midwestern metropolitan area and its three middle-class suburbs, with a population of over 200,000 residents. Phone numbers were randomly selected through a systematic sampling procedure from the local phone listings.

The Sample

Of the 516 adults who responded to the survey (for a 74% response rate), 53.3% or 233 were VCR owners. Only the results gathered from the 233 VCR owners are considered in this study. In terms of video-technology availability among these VCR homes, 82.3% had basic cable, 49.7% were pay-cable subscribers, and 11.7% owned a videocamera. The demographic composition for the sample indicated that the average VCR owners were between 25 and 45 years of age, college-educated, and had an annual income of $33,000 to $37,000; the majority were married and had children. In terms of gender distribution, the sample was evenly split between males and females. With regard to time devoted to different VCR-use purposes (with 100% being the total), an average user utilized the VCR for: (a) time-shifting 30.7% of the time, (b) building a video library 8.6% of the time, (c) watching pre-recorded tapes 47.9% of the time, and (d) playing back home movies 3.6% of the time. Finally, these VCR users reported watching 3.2 hours of television on average during a typical day.

Measurement

The audience activity construct was assessed by a number of viewing-related activities reflecting the selectivity, involvement and utility from the pre-exposure, during exposure, and post-exposure phases. These variables were primarily based on existing work (Blumler, 1979; Levy & Windahl, 1985; Levy, 1987), and were all measured on a 5-point Likert scale (i.e., "very often," "often," "sometimes," "rarely," and "never").

For *selectivity* before exposure, respondents were asked how often they made viewing plans, used program guides to help make viewing decisions, and felt they had a lot to choose from. Items assessing *involvement* during exposure included: how often respondents zapped commercials, switched channels to screen other viewing options, reselected a viewing choice during viewing, and watched more than one program at a time. The *utility* dimension during exposure was measured by how often respondents discussed the program content under exposure with co-viewers. The three questions comprising *utility* after exposure were intended to reveal how often respondents: (a) engaged themselves in certain activities due to the influence of a TV announcement or advertisement, (b) made product purchases due to the influence of TV advertisements, and (c) discussed program content with other people after exposure.

Two additional viewing-related variables were studied. One of them measured the level of viewing satisfaction by a 4-point scale (i.e., "very satisfied," "satisfied," "unsatisfied," and "very unsatisfied," in a descending order). The other reflected the number of hours spent in watching TV on a typical weekday and weekend; the average of these two categories was computed to obtain the average daily TV viewing time.

The VCR-use activity construct was composed of recording and replaying among other related activities. Recording activity was measured by asking the respondent how many programs he or she recorded from pay cable, basic cable, and broadcast network channels in a typical week. A recording-related activity, viewing activity during recording, was determined by the percentage of time respondents spent watching the channel being recorded, watching the channel not being recorded, and not watching TV during recording. The one recording-related audience activity, *commercial pausing* (i.e., pausing commercials during recording) was measured by the same 5-point Likert scale described earlier.

Replaying activity was assessed by how many programs respondents recorded from pay cable, basic cable, and broadcast network channels that were played back in a typical week. There were two playback-related activities specified: (a) days elapsed before playbacks, ranging from 0 to 7 or more days, and (b) time of day for playbacks, measured in 2-hour intervals, starting from 8a.m. until after 11p.m. The only playback-related audience activity, commercial zipping (i.e., fast forwarding commercials during playbacks), was measured by the same 5-point Likert scale just described.

Two other VCR-use related variables were also examined. *VCR-use satisfaction,* which reflected the level of reported satisfaction with VCR use, was measured by a 4-point scale (i.e., "very satisfied," "satisfied," "unsatisfied," and "very unsatisfied," in a descending order). *TV-viewing quality,* which assessed whether VCR use improved respondents' TV viewing quality, was measured by a 4-point scale (i.e., "greatly improved," "improved," "improved a little," and "not improved," in a descending order).

Data Analysis

Frequency distributions were provided for recording and replaying activities as well as for other recording and replaying related activities such as TV viewing activity during recording, days elapsed before playbacks, and time of day for playbacks. Each measure for the audience activity construct, including commercial pausing and commercial zipping, were descriptively summarized; the mean value for each variable was also computed. Finally, Pearson product moment coefficients were obtained to show the correlations between VCR-use activity and general audience activity measures, as well as interrelations among various VCR-use activity measures.

FINDINGS

Across the three different recording sources, broadcast network programs reportedly were the most frequently recorded and replayed; basic cable programs ranked second, followed by pay-cable programs (Table 4.1). Although a majority of VCR users recorded 1–3 programs per week, less than one-fifth recorded 4–6 programs, less than 2% recorded 7–9 programs, and approximately 8% recorded 10 or more programs. Overall, 66.5% of all recordings made were played back, and the average numbers for weekly recording and playbacks per VCR user were 3.2 and 2.6 copies, respectively.

In terms of viewing activities during recording (with 100% being the total), an average VCR user turned off the TV set 48% of the time,

TABLE 4.1
Weekly Recording and Replaying Activities

	PAY CABLE		*BASIC CABLE*		*BROADCAST NETWORKS*		*TOTAL*	
	# of programs	*n*	*# of programs*	*n*	*# of programs*	*n*	*# of programs*	*n*
1–3 Programs								
Recorded %	51.9	83.0	35.6	74.7	31.2	67.9	35.6	73.1
Replayed %	71.0	92.3	55.6	83.6	32.0	68.5	43.7	78.3
Playback %	87.5	92.3	85.1	91.1	76.6	80.3	81.6	86.5
4–6 Programs								
Recorded %	22.2	10.6	22.5	16.0	27.4	20.5	25.1	17.1
Replayed %	17.4	5.1	24.1	11.5	27.6	20.2	25.3	14.3
Playback %	50.0	40.0	58.2	58.3	75.2	70.3	67.0	67.5
7–9 Programs								
Recorded %	7.4	2.1	3.0	1.3	3.7	1.8	4.0	1.7
Replayed %	11.6	2.6	5.3	1.6	5.0	2.2	6.0	2.1
Playback %	100	100	100	100	100	100	100	100
10 or More Programs								
Recorded %	18.5	4.3	38.9	8.0	37.7	9.8	35.3	8.1
Replayed %	0.0	0.0	15.0	3.3	35.4	9.0	25.0	5.3
Playback %	0.0	0.0	21.1	33.0	70.0	72.7	47.2	52.6
Total								
Recorded	108	47	244	75	398	112	750	234
row %	14.4	20.1	32.5	32.0	50.1	47.9		
Replayed	69	39	133	61	297	89	499	189
row %	13.8	20.6	26.7	32.3	59.5	47.1		
Playback %	63.9	82.9	54.5	81.3	74.6	79.5	66.5	56.6
X Recordings	2.3		3.3		3.6		3.2	
X Playbacks	1.8		2.2		3.3		2.6	

TABLE 4.2
TV Viewing Activity during Program Recording

	%	*n*
Watching the Same Channel	28.3	232
Watching Another Channel	23.7	
Not Watching Television	48.0	

watched a different program not being recorded 23.7% of the time, and watched the same channel being recorded 28.3% of the time (Table 4.2). With regard to playback-related activity (Table 4.3), it appeared that the majority of VCR users played back their recordings within the next 2 days, nearly 13% of them played back tapes between the third and sixth days after recording them, and another 13% did so on the seventh day; the average number of days elapsed before playbacks was 2.4 days. Another playback-related activity showed the time of day VCR users generally viewed what they had taped (Table 4.4). It seems that at least 50% of VCR homes played back their recordings between 7p.m. and 11p.m., or during the fringe and prime-time broadcast hours; another 10%–11% did the same within the early fringe of 5p.m. to 7p.m., and after 11p.m., the late night television slot.

Table 4.5 provides the description of VCR users' involvement with various audience activities, including a recording-related and playback-related activity. The majority of this sample reported planning their viewing and program selections "very often" or "often," indicating a high degree of selectivity during the pre-exposure period.

In terms of during-exposure involvement, nearly 50%–60% of the audience members were often behaviorally involved with the viewing process in that they zapped commercials, switched around channels, and

TABLE 4.3
Days Elapsed before Playing Back Recording

Days	%	$\overline{X}$	*n*
0	18.7	2.4	182
1	34.1		
2	17.0		
3	7.7		
4	3.8		
5	1.1		
6	0.5		
7	13.7		
9	3.3		

TABLE 4.4
Time of Day for Playbacks

Time	%	*n*
8–10 a.m.	2.6	231
10–noon	3.5	
1–3 p.m.	4.8	
3–5 p.m.	7.8	
5–7 p.m.	11.3	
7–9 p.m.	31.2	
9–11 p.m.	19.9	
After 11 p.m.	10.4	

TABLE 4.5
Frequencies for the Occurrence of Audience Activities

	Very Often (%)	*Often* (%)	*Sometimes* (%)	*Rarely* (%)	*Never* (%)	$\overline{X}$	*n*
Commercial pausing	39.4	20.6	2.4	11.2	26.5	3.4	170
Commercial zipping	49.7	19.5	8.1	8.6	14.1	3.8	185
Viewing planning	27.0	35.7	15.7	15.2	6.5	3.6	230
Program-guide use	32.6	27.0	18.3	15.2	7.0	3.6	230
Program choices	18.3	35.7	26.1	17.8	2.2	3.5	230
Commercial zapping	27.6	22.8	12.3	19.3	18.0	3.3	228
Channel switching	28.8	32.8	17.9	14.0	6.6	3.6	229
Channel reselecting	24.1	33.3	20.6	16.2	5.7	3.5	228
Multichannel viewing	7.9	11.8	13.5	26.6	40.2	2.2	229
During-viewing discussion	7.0	35.5	28.5	20.2	8.8	3.1	228
Post-viewing discussion	10.9	43.2	31.0	13.1	1.7	3.5	229
Post-viewing activity	1.3	6.6	28.1	51.3	12.7	2.3	228
Post-viewing purchase	0.0	7.9	31.0	46.3	14.8	2.3	229

Very Often = 5
Often = 4
Sometimes = 3
Rarely = 2
Never = 1

reselected viewing choices; approximately 20% of them often watched more than one channel at a time. These types of active involvement with the viewing process were well paralleled by the similar behavioral involvement occurring during the recording and replaying of TV programs, as at least 60% of the audience reported pausing commercials during recording and zipping commercials during playbacks.

The interpersonal communication utility was apparently "often" realized by more than 40% of the sample during viewing and by more than 54% after viewing. However, the majority of the sample did not utilize information such as TV announcements or advertisements for initiating certain activities or purchasing products, although between 28% and 31% did "sometimes" take advantage of those utility functions of TV viewing activity.

Table 4.6 demonstrates the intercorrelations between VCR-use related activities and general viewing-related audience activities. It appeared that commercial zapping was significantly related to all VCR-use related activities, signifying an extensive use of remote-control devices among VCR users and an active level of attempting to control their viewing conditions.

Although more active recorders tended to be more selective during the pre-exposure period, they also absorbed media information more frequently for personal utilitarian activities. Interestingly enough, viewers who played back more recordings were rather selective before exposure (e.g., checked the program guide to avoid missing a favorite program) and less involved during exposure (i.e., through viewing more than one channel at a time).

Viewers who often paused commercials during viewing were more involved in improving their viewing choices and more heavily influenced by commercial messages, as a result of inadvertently watching more commercials or paying greater attention to them. By contrast, viewers who often fast-forwarded commercials during playbacks watched a lot less TV and were not involved with any audience activities except for commercial zapping.

Those for whom TV-viewing quality was better improved due to VCR use were more selective before exposure, more involved with improving viewing choices, and more capable of capitalizing on the interpersonal communication utility during exposure. These viewers also recorded and played back more recordings, and were generally more satisfied as well as heavier TV viewers. Despite the expectation of an intercorrelating nature between VCR-use satisfaction and a variety of different audience activities, few significant relations were found. The more satisfied VCR users were simply those who appear more active in zapping commercials, making and replaying recordings, and those who consider their TV-viewing quality improved (due to VCR use).

TABLE 4.6
Correlations between VCR-Use Activities and Audience Activities

	Weekly Recordings	*Weekly Playbacks*	*Commercial Pausing*	*Commercial Zipping*	*TV-Viewing Quality*	*VCR-Use Satisfaction*
Viewing planning	.18*	.20*	.04	.08	.16*	.00
Program guide use	.10	.13*	−.06	.03	.06	.05
Program choice	.09	.06	−.09	.09	.09	.17
Commercial zapping	.13*	.15*	.13*	.16*	.13*	.14*
Channel switching	.04	.09	.11	.01	.11	.04
Channel Reselecting	.02	.08	.13*	.06	.21*	.04
Multichannel viewing	.10	.14*	.09	.05	.11	.04
During-viewing discussion	.06	.01	.09	−.01	.13*	.02
Post-viewing discussion	.09	.01	.03	.06	.07	.00
Post-viewing activity	.19*	.03	.10	−.08	.07	−.01
Post-viewing purchase	.14*	.03	.14*	−.06	.09	.00
TV-viewing satisfaction	.07	.10	.00	−.01	.16*	.09
TV-viewing time	.07	.04	.08	−.20*	.13*	.06
Weekly playbacks	.59*	—	—	—	—	—
Commercial pausing	.11	−.03	—	—	—	—
Commercial zipping	.07	.15*	.15*	—	—	—
TV-viewing quality	.16*	.16*	.07	.09	—	—
VCR-use satisfaction	.13*	.13*	.09	.07	.24*	—

DISCUSSION

Compared to Levy's (1983) earlier study, it is apparent that program recording and replaying activities for this group of VCR users are less intensive. In particular, only 56.6% of the viewers played back 66.5% of all recordings made (Table 4.1), which is a revelation that time-shifting does not guarantee the intended delayed viewing. Of course, the differences in the results of these two studies could also be due to the different research methods used. Levy's study collected data through a 1-week diary approach, whereas the present study was a telephone survey. Although diary data often suffer from validity problems, telephone surveys also may present potential validity problems if respondents cannot accurately recall their media-use behavior. Therefore, cautions should be taken into account when self-reported data such as the present study are discussed.

There are perhaps a variety of reasons why recorded programs are not played back. When playbacks do take place, the process often requires preempting the viewing of certain TV programs or other activities. As reported in Table 4.4, more than 50% of the VCR users played back previously recorded programs between 7p.m. and 11p.m. Although more than 60% of American TV homes have at least two TV sets, it is difficult to know what percent of viewers' evening activities (whether they are viewing related or nonviewing related) during those hours are altered due to TV program playbacks.

Time-shifting activities have now declined to the point where surging video rental viewing recently surpassed the time-shifting usage frequencies (A. C. Nielsen, 1988). Low rental costs and an abundant variety of video titles, compounded with the narrowing of theatrical and video release windows to three months or less, signal a steady increase in video rental activity (see Komiya and Litman, chapter 2, this volume). An additional reason, as mentioned earlier in this chapter (*Cablevision,* 1985), is that some VCR owners are incapable of interfacing the VCR with a TV set. Dobrow (1987) also found that some VCR users are not able to program the VCR for time-shifting purposes.

Regardless of the degree of alteration in a viewer's regular TV viewing schedule, due to time-shifting or video-renting activities, the concept of prime-time TV has become questionable. According to recent ratings statistics, VCR use has been steadily eroding broadcast program ratings by segmenting the audience shares (A. C. Nielsen, 1989; National Association of Broadcasters, 1989). Active use of the VCR medium seems to imply an ever-growing trend of home entertainment or the formation of a home video culture. In fact, according to one recent study (Lin, 1988), after acquiring a VCR, more than half of VCR owners stated that they watched more TV at home and spent more time watching videos instead of theatrical films. These findings reflected an almost 50% increase for the similar findings reported by an earlier study (Harvey & Rothe, 1986).

If VCR owners have become more "home-bound" due to VCR usage, this phenomenon is indicative of TV viewers' desire to enjoy more diverse "home video" contents. As in the case with magazine and book dissemination, videos are striving to accommodate divergently interested viewers by offering a diverse range of titles that cover such areas as entertainment, history, science, nature, instructional lessons, as well as various hobbies (e.g., sports). It appears that TV viewing no longer stands only for watching TV programs provided by TV programmers; instead, it may suggest watching a special-interest video, such as gardening guide through the use of a VCR and TV monitor.

By implication, the concept of "TV viewer" or "TV audience" perhaps needs reexamination in terms of media content consumed in conjunction with the audience activity performed and video technologies adopted by an audience member. Within the limitations of the present study, a majority of VCR owners were found to have actively engaged in most audience activities measured from the pre-exposure through post-exposure periods.

In particular, the average audience member was often "selective" in making a viewing decision. Such a selective process was related to the frequencies of "plans" to alter regular TV viewing schedules. Because the time-shifting enthusiasts (who can record programs for themselves and/or others) and playback faithfuls may not be the identical group of viewers, they behave somewhat differently. The former often planned ahead what to watch and record with or without the use of a program guide, perhaps due to their familiarity with the schedules of regularly recorded programs. By contrast, the latter often used a program guide to plan ahead what to watch and/or when to play back, so that a more optimal playback schedule could be determined without interfering with the viewing of other programs.

The average VCR users in this sample were "sometimes" to "often" "involved" with improving their viewing conditions during exposure by switching channels to skip commercials, explore alternative viewing options, and reselect a viewing choice. Such behavioral involvement with manipulating one's viewing condition is not contingent on VCR-use activities, however. Access to a remote-control device (which usually comes as a standard piece of equipment with cable subscription and VCR ownership) elicited the ultimate "active" inclination in viewers, who then took advantage of the technical capability to enhance "control" over their viewing process. By implication, increased channel switching activity points to a less loyal audience who is selectively involved with their desirable programs and highly aware of their viewing preferences (Heeter, 1986; Heeter & Greenberg, 1985).

Although many VCR users were capable of utilizing their viewing experience for interpersonal communication purposes during and post-viewing exposure, such a utilitarian approach implies an active audience that is cognitively and affectively as well as behaviorally involved with the program content. Although overall viewing attention may decline due to program discussion during exposure (Levy & Windhal, 1985), viewers simply transfer part of this cognitive processing energy to discussion activity, which requires an even higher level of cognitive involvement. The lack of any correlation between both viewing discussion and most VCR use measures suggests that viewers were actively involved in program discussion whether or not they were active VCR users. Nevertheless, the significant correlation between during-viewing discussion and improved TV-viewing quality (due to VCR ownership) does reveal that, with a VCR, time-shifting has made more family or group viewing possible and thus facilitates more interpersonal communications (Levy, 1987; Rubin & Bantz, 1987). However, such an observation did not hold true for Gunter and Levy's (1987) findings, which indicated that VCR use did not increase the quantity of time spent with others among a group of British respondents; instead, content preference seemed to dictate more individualistic VCR use.

Rather less common than interpersonal discussions are post-exposure activities such as post-viewing activity and post-viewing purchase activities, which involve most viewers only rarely, if at all. However, viewers who more frequently recorded and played back programs did appear to absorb more information presented in TV advertisements or announcements for personal activities (e.g., sports events, charity functions) and product purchases. By contrast, viewers who were only engaged in playback but not recording activity did not seem to learn and utilize the TV transmitted information for carrying out personal activities and product purchases. Perhaps the former represented the least obstinate or serious viewers, who more often perceived TV-transmitted information as believable or acceptable, and thus were more impressionable in terms of undertaking activities portrayed on TV. This may occur despite the association between post-viewing purchases and commercial pausing, as such censorship requires the viewer to pay close attention to the commercial messages being paused from recording. Thus, even when commercials are "paused" off the final tape, those making recordings have already paid close attention to the commercials being skipped.

CONCLUSION

This home video culture is perhaps part of the overall progression of a visual or video culture, one strengthened by the innovation and adoption of various telecommunication technologies such as video phones, video conferencing, and computer networking, among others. An important characteristic of this video culture is that the distinction between mass communication and point-to-point communication seems to blur gradually. For instance, the video conference is considered a point-to-point mode of communication. However, through the use of satellite technology, such a video conference can also connect a mass of people at various locations around the world. In the case of two-way cable communication, when viewer feedback is transmitted to the cable headend, the transmission is apparently done for the purpose of point-to-point communication through a mass communication channel.

In similar fashion, VCRs may transform TV from a mass to an individually or perhaps family-oriented medium. And, because VCRs are the most widely diffused of the new technologies, most people are getting their first experience with enhanced audience control through that medium. By implication, VCRs have made the creation of idiosyncratic home video or media subcultures possible, by giving audience members the opportunity to shape their own home video or media environment. As suggested by the evidence reported in the present study, VCR owners enjoyed being active users rather than passive viewers by utilizing their TV and VCR set as tools to program their own home entertainment fare and to meet their demands for visual informational learning. Further research should continue to examine how these viewers make use of other VCR capabilities in areas such as computer interfacing, in order to gain a better understanding of audience behavior in future multimedia environments.

REFERENCES

A. C. Nielsen. (1988). *Nielsen research report.* Northbrook, IL: Author.

A. C. Nielsen. (1989). *Nielsen report on television.* Northbrook, IL: Author.

Bauer, R. A. (1964). The obstinate audience: the influence process from the point of view of social communication. *American Psychologist, 19,* 314–328.

Blumler, J. G. (1979). The role of theory in uses and gratifications studies. *Communication Research, 6,* 9–36.

Bogart, L. (1965). The mass media and the blue collar worker. In A. Bennet & W. Gomberg (Eds.), *The blue collar-world.* Englewood Cliffs, NJ: Prentice-Hall.

Dobrow, J. R. (1987). *The social and cultural implications of the VCR: how VCR use concentrates and diversifies viewing.* Unpublished doctoral dissertation, University of Pennsylvania, Philadelphia, PA.

Donohue, T., & Henke, L. (1988). *The impact of videocassette recorders on traditional television and cable viewing habit and preferences.* Research Report, National Association of Broadcasters, Amherst, MA.

Galloway, J. J., & Meek, F. L. (1981). Audience uses and gratifications: An expectancy model. *Communication Research, 8,* 135–450.

Greenberg, B. S., & Lin, C. A. (1989). Adolescents and the VCR boom: Old, new and non-users. In M. Levy (Ed.), *The VCR age: Home video recorders and the mass communication process.* Beverly Hills, CA: Sage.

Greenwald, A. G., & Leavitt, C. (1984). Audience involvement in advertising: Four levels. *Journal of Consumer Research, 11,* 581–592.

Group W cable makes peace with VCRs. (1985, February 25). *Advertising Age.*

Gunter, B., & Levy, M. R. (1987). Social contexts of video use. *American Behavioral Scientist, 30,* 486–494.

Harvey, M. G., & Rothe, J. T. (1986). Video cassette recorders: Their impact on viewers and advertisers. *Journal of Advertising Research, 25,* 19–27.

Heeter, C. (1986). Program selection with abundance of choice: A process model. *Human Communication Research, 12,* 126–152.

Heeter, C., & Greenberg, B. S. (1985). Profiling the zappers. *Journal of Advertising Research, 25,* 15–19.

Kellerman, K. (1985). Memory processes in media effects. *Communication Research, 12,* 83–131.

Lemish, D. (1985). Soap opera viewing in college: A naturalistic inquiry. *Journal of Broadcasting & Electronic Media, 29,* 275–93.

Levy, M. R. (1978). The audience experience with TV news. *Journalism Monographs, 5.*

Levy, M. R. (1983). Conceptualizing and measuring aspects of audience activity. *Journalism Quarterly, 60,* 109–114.

Levy, M. R. (1987). VCR use and the concept of audience activity. *Communication Quarterly, 35*(3), 267–275.

Levy, M. R., & Fink, E. L. (1984, Spring). Home video recorders and the transience of broadcasts. *Journal of Communication,* pp. 56–71.

Levy, M. R., & Windahl, S. (1985). The concept of audience activity. In K. E. Rosengren, L. A. Wenner, & P. Palmgreen (Eds.), *Media gratifications research: Current perspectives* (pp. 109–122). Newbury Park, CA: Sage.

Lin, C. A. (1987). A quantitative analysis of worldwide VCR penetration. *Communications, 13*(3), 131–159.

Lin, C. A. (1988, July). *Assessing the impact of the evolution of home video culture.* Paper presented to the annual conference, Association for Education in Journalism and Mass Communication, Portland, OR.

McGuire, W. J. (1974). Psychological motives and communication gratification. In J. G. Blumler & E. Katz (Eds.), *Current perspectives on gratifications research.* Beverly Hills, CA: Sage.

MSOs position cable as VCR friendly. (1985, February 26). *Cablevision,* p. 16.

National Association of Broadcasters (1989, January 30). *TV today.*

Palmgreen, P., & Rayburn, J. D. (1979). Uses and gratifications and exposure to public TV. *Communication Research, 7,* 162–192.

Rubin, A. M., & Bantz, C. R. (1987). Utility of videocassette recorders. *American Behavioral Scientist, 30,* 167–185.

Rubin, A. M., & McHugh, M. P. (1987). Development of parasocial interaction relationships. *Journal of Broadcasting & Electronic Media, 31,* 279–92.

Rubin, A. M., & Perse E. (1987). Audience activity and soap opera involvement: A uses and effects investigation. *Human Communication Research, 14,* 246–268.

Sony begins roll of videocassette units in N.Y. market. (1975, November 3). *Advertising Age.*

Standford, S. W. (1984). Predicting favorite TV program gratifications from general orientations. *Communication Research, 11,* 519–536.

Wenner, L. (1985). Transaction and media gratifications research. In K. E. Rosengren, L. A. Wenner, & P. Palmgreen (Eds.), *Media gratifications research: Current perspectives* (pp. 241–252). Newbury Park, CA: Sage.

Yorke, D. A., & Kitchen, P. J. (1985). Channel flickers and video speeders. *Journal of Advertising Research, 25,* 21–27.

5

VCRs and People's Control of Their Leisure Time

Kimberly K. Massey
University of Utah
Stanley J. Baran
San Jose State University

> *Between the end of 1987 and 1988, scientists added one second to the world's clock. It wasn't enough. "I'd like to but I don't have the time" is rapidly replacing "have a good day" as the signature line of the later 80's. In addition, stress has taken the place of paranoia as the mental affliction of the generation.*
>
> —Richard (1988)

Time seems to be on everybody's mind. As work becomes less industrial and more service-oriented, it tends to revolve around time—meetings and deadlines are the order of the business day. Travel arrangements must now be made at least 30 days in advance to get the best prices. FAX machines and overnight mail services have done away with lag time (the safety cushion of the tardy) between project completion and actual delivery. Time must be planned and facilities reserved for health and recreational leisure activities such as golf and racquetball. Even sleep is structured into our busy days. Today, mastery of scheduling has become necessary to keep pace in the rat race.

Adhering to the rule of supply and demand, the American people are willing to pay—and pay dearly—to get and save time. Leisure time has become an almost priceless commodity. "Time could end up being to the '90s what money was to the '80s" (Gibbs, 1989, p. 58). Savvy business minds hoping to capitalize on the time drought have invented numerous ways to help the public ease the crunch. Services can be hired to do the shopping for us out of a catalogue, pick up our dry cleaning, or cater dinner for us and our friends. Greeting cards written specifically for children with slogans varying from "Wish I were there to tuck you in" to "I will miss you today" are being marketed for busy parents to compensate for those lost "precious moments." Three-minute microwave entrees have replaced what we formerly thought of as time-saving 30-minute TV dinners and our automobiles have become offices away from work complete with telephones, lap-top computers and portable FAX machines.

Americans are running out of time, or so it seems. According to Nancy Gibbs (1989), a poll for *Time* and CNN conducted by Yankelovich Clancy Schulman described this sense of "not having enough time" as being "especially acute among women in two-income families [an estimated 57% of American households are two-income]: 73% of the women complain of having too little leisure, as do 51% of the men" (p. 61). Gibbs also assured us that the pressure we feel from the clock is certainly not a figment of our imaginations. She wrote:

> According to a Louis Harris Survey, the amount of leisure time enjoyed by the average American has shrunk 37% since 1973 [from 16.6 hours per week to 9.6]. Over the same period, the average workweek, including commuting, has jumped from under 41 hours to nearly 47 hours. In some professions, predictably law, finance and medicine, the demands often stretch to 80-plus hours a week. Vacations have shortened to the point where they are frequently no more than long weekends. And the Sabbath is for—what else? shopping. (p. 59)

John Robinson, Director of the Americans' Use of Time Project, disagreed, at least in part. He wrote:

> Americans have more free time today than ever before. Men have 40 hours of free time a week, and women have 39 hours. Free time is defined as what's left over after subtracting the time people spend working and commuting to work, taking care of their families, doing housework, shopping, sleeping, eating and doing other personal care activities. (1989, p. 34)

In his analysis, this free time includes such activities as adult schooling, club and other organizational activities, sports, recreation and hobby activities, reading, visiting friends and relatives, and watching television. And although his data indicate that the "40-hour work week is balanced by a 40-hour play week," he conceded that many of us are under a real time crunch. Working parents are especially beleaguered and the "baby boomers," those between the ages of 36 and 50, have proportionately less free time than do other age categories.

Leisure time may or may not be hard to come by, but it is difficult to define and even harder to measure. In fact, the discrepancies between the Harris results on which Gibbs reported and the Robinson analysis may be due in part to method (survey vs. diary) and, in part, due to definition. Robinson (personal communication, July 13, 1989) himself has substituted the term "free time" for the word "leisure" in his research. In his seminal 1969 work on "television and leisure time" (reporting on 1965 data), for example, he defined leisure as "all daily activities outside of work, housework, child care, shopping, sleep, and other personal care" (p. 271). In 1981, when he wrote on his 1975 replication of that work, he still used the word "leisure" and employed "the same coding procedures for activities . . . in both studies" (p. 123). This older, twice-utilized definition varies very little from that employed in his 1989 work. Why, then, the shift to "free time?" Because so much of our free (nonwork, nonmaintenance) time is filled with so many options and activities, and these options and activities have become so crucial to our mental and physical health as well as to our conceptions of ourselves that free-time often hardly seems like leisure time. Quoted in Gibbs (1989, p. 59), Robinson said, "People's schedules are more ambitious. There just isn't enough time to fit in all the things one feels have to be done." Adult schooling, for example, may consume free time, but if taken seriously, it may not be very leisurely.

Rojek (1985) used the adage that what is "work for some is leisure for others" to discuss the problems social scientists have had in defining the field, yet the majority of observers choose to equate leisure and free time. Recognizing that a precise definition of leisure time is needed for operationalization in research, he ultimately adopted Vickerman's rendition of leisure (and so do we in this chapter): "take leisure time be roughly equivalent to free time; that time left over after meeting commitments to work and such essential human capital maintenance as sleeping, eating, and personal hygiene" (p. 14).

LEISURE AS A CONCEPT

The concept of leisure did not receive much scientific attention until soon after the Second World War. Since that time a number of factors have combined to increase the importance and the amount of leisure time available to individuals and to create new options for filling that free time:

1. Industrialization, urbanization, and automation have caused a "natural shift" from a work-centered to a leisure-centered life.

2. The public experienced a growth in disposable income. This is especially true for younger people.
3. More employers began experimenting with 4-day work weeks and computerization.
4. In 1971 a Federal law shifted 5 midweek holidays to positions adjacent to weekends, providing a set of 4-day weeks for a majority of the U.S. work force.
5. Improved pension plans have allowed for earlier retirements.
6. Advances in science have resulted in longer life spans.
7. The scale and range of leisure goods and services have expanded.
8. Women are doing less housework than they did in "pre-appliance" days and before they were successful in enlightening their male domestic partners to the merits of labor sharing at home.
9. Fewer households have children (freeing people from the time demands of child care), and couples are marrying later, therefore spending more of their lives unmarried; single people have more free time than do married folks. (See Harvey & Rothe, 1985; Nayman, Atkins, & Gillette, 1973; Robinson, 1989; Rojek, 1985; Tinsley, Barrett, & Kass, 1977 for details.)

Although the contemporary time crunch that many of us feel may make it hard to believe, these events, most related to war-driven or war-accelerated alterations in our culture, technology, and demographics, produced a sudden and unprecedented burst of leisure time and, as a result, scientific interest. Social scientists began to focus on the "progressive" phenomenon of leisure, often considering it "real evidence of free choice and personal liberty in the western democracies" (Rojek, 1985, p. 2). Researchers (e.g., Tinsley et al., 1977) began to connect leisure activity choice with an individual's satisfaction predicting that, "as leisure time increases, the life satisfaction of an individual will become increasingly dependent upon the extent to which that person is able to select leisure activities which fulfill his or her needs" (p. 111).

TELEVISION AND LEISURE

The changes in America just enumerated occurred simultaneously with a significant change in the American media environment—namely, the widespread introduction and diffusion of television.

When the Television Freeze of 1948 was lifted in 1952, there were fewer than 250,000 television receivers in the United States. By 1960, a mere 8 years later, 80% of all American homes had at least one set (Baran, McIntyre, & Meyer, 1984). Because the network structure, recognizable and popular stars, and means of economic support were transferred wholesale from radio to the infant medium, people almost instantly found themselves with a new and improved method of enjoying the leisure-time bonanza. Examining the steady rise in Nielsen viewing figures, which ultimately reached and have remained constant for the last several years at approximately 7 hours a day for an average household, there can be little doubt that television viewing did in fact become a common leisure pastime.

Robinson (1969) identified this phenomenon by writing,

> Television has had a massive impact on American daily life, responsible for a greater rearrangement of time usage than the automobile. Furthermore, the time now devoted to television is of such magnitude that it has apparently not only usurped time previously devoted to other mass media, but has eaten into substantial portions of time previously spent in other forms of leisure. (p. 211)

His analysis of the 1965 diaries indicated that "over the whole population close to 28% of all leisure time appears to be spent primarily watching television" (p. 213).

In reporting on the 1975 diaries, Robinson (1981) argued that several factors should have suggested no or limited increases in the amount of leisure devoted to television. In the span between his two investigations, the number of hours devoted to the average work week failed to decline; so the amount of available nonleisure time remained static. More women entered the workforce, leaving them less time for viewing. The adult population became better educated, and a higher level of education was the single best predictor of smaller amounts of television consumption at that time. Personal income and, therefore, the number of other leisure-time options grew. Increased movement of the population to the Sun Belt meant that more people had additional outside-the-home leisure options. The country's romance with "expressive personal growth activities" would have suggested less attention to the passive, unimaginative activity of television viewing. Finally, the novelty of television may well have declined in those 10 early years of the medium's life. So, of course, he found significant *increases* in the amount of leisure time devoted to viewing across all demographic groups in the country. In fact, television viewing now accounted for 40% of all leisure-time activity. He attributed the increase to two factors: (a) improvements in programming content due to the programmers' and advertisers' desire to reach the better educated and heavier spending people in their audiences; and (b) improvements in technology, specifically color television and cable.

VIDEOCASSETTE RECORDERS AND LEISURE

Now, another technology, the VCR, has entered into our leisure-time equations. It not only occupies and enriches individuals' free time, but it has directly affected another leisure habit—television viewing. Concentrating on the nature and capabilities of VCR technology alone, however, may be a wasted exercise when striving for an understanding of its impact on leisure. Instead, following the advice of Marvin (1988), we should leave behind the traditional notions that new technologies fashion new audiences out of "voiceless collectives and inspire them to new uses based on novel technological properties." Rather, she suggested that emerging technologies provide a new stage for existing groups to "negotiate power, authority, representation, and knowledge with whatever resources are available" (p. 7).

When applying these ideas more specifically to media technology she continued,

> Media are not fixed natural objects; they have no natural edges. They are constructed complexes of habits, beliefs, and procedures embedded in elaborate cultural codes of communication. The history of media is never more or less than the history of their uses, which always lead us away from them to the social practices and conflicts they illuminate. New media, broadly understood to include the use of new communication technology for old or new purposes, new ways of using old technologies [what VCRs do to television] and, in principle, all other possibilities for the exchange of social meaning, are always introduced into a pattern of tension created by the coexistence of old and new, which is far richer than any single medium that becomes a focus of interest because it is novel. (p. 8)

In the case of the VCR, she might argue, a most valuable means of understanding its impact is to examine this technology's alterations in how we use television and how that use has affected various aspects of our environment. Leisure is one of those aspects. Although not speaking exclusively of the VCR, Hornik and Schlinger (1981) labeled this more holistic view as the analysis of "the consumption of the media situation." That is, media technologies interacting with one another and, in combination, with the consumption environment.

From its inception, broadcasting's critics predicted that the videotape recorder would one day accompany television sets in every living room, creating Marvin's hypothetical "new audiences." DeLuca, for example, proposed that this accomplishment "would free viewers from their total reliance on broadcasting and in particular on the dominant commercial networks with their monotonous mass-taste programming" because economically "a network could not consider programming for a prospective audience of say 100,000 people nationwide, a producer of prerecorded video tapes could reap a bonanza from a single program that sold 100,000 copies" (DeLuca, 1980, p. 85). In other words, the VCR was to become a new medium with new audiences.

On the other hand, broadcasters were initially unconcerned about the use of VCRs. They did not care *when* people watched their shows as long as they *did* watch them. In other words, *when* audiences watched might change a bit, but not *what* and *how* they watched. Marvin would say they underestimated the "pattern of tension" created by the coexistence of the old and new; and, as Secunda (chapter 1, this volume) confirms, both industry insiders and critics were wrong.

Critics, broadcasters, theater owners, film producers, and commercial sponsors are now all paying close attention to what people are doing with their VCRs, because even though the VCR was initially conceived of as a complement to the television set, the *way* in which people are using the VCR in their viewing is changing their leisure and therefore, the television industry that for so long has prospered by successfully filling that free time. In Marvin's parlance, the VCR *and* television set form a new leisure use technology "which is far richer" than the old (television) or the new (VCR) taken alone.

The "power and authority" granted to television viewers by the VCR that impact free time most obviously include:

1. *Zapping,* where viewers edit out commercials while recording programs with their remote controls and *zipping,* where viewers scan through the advertisements on taped programs at high speeds. Both grant viewers more control over how they spend their leisure time and it increases the available amount of that time. Video industry data indicate that nearly half the households that have a remote engage in this "video grazing."
2. *Time-shifting,* where television programs are recorded off-channel on the VCR, permitting viewers to watch programs when it is most convenient (more control over leisure), resolving programming conflicts (more enjoyable leisure) and allowing previewing of programs for video libraries or for children's consumption (more rewarding use of leisure). Moreover, Nielsen figures indicate that 67% of all off-station taping is devoted to television network fare, suggesting audience "prioritizing" of leisure-time viewing. Their data also show that

50% of all home taping occurs while the television set is turned off, strongly indicating that the audience is doing something other than viewing at the time; that is, in many cases they are prioritizing their leisure activities (Nielsen Media Research, 1988).

3. Establishing a controlled environment for children. VCRs provide parents an opportunity to influence the viewing behavior of their children (see Jordan, chapter 9, this volume). The controversy over advertising to children, for example, is well documented. By "playing prerecorded programs or programs recorded by the parent (with commercials deleted) [parents can] shelter children from advertising messages" (Harvey & Rothe, 1985, p. 20). However, other studies of parents and children indicate that children do not perceive the same amount of control that parents claim, and that parents often use the VCR as just more TV (see Kim, Baran, & Massey, 1988).

4. Increasing the amount of noncommercial television viewing. Viewers can now accumulate tape libraries and rent prerecorded tapes, permitting them to watch more noncommercial television. This not only allows for greater personal discretion in how free time is spent, but many critics argue that the result is an improvement of commercial television programming as the networks strain to compete with various other VCR program options. This echoes Robinson's (1981) argument that broadcasters maintained and even expanded the amount of viewing between the years of 1965 and 1975 through improved programming even though many factors would seem to have encouraged lower amounts of television consumption.

5. Faster viewing of programs. The fast-forward mechanism permits viewers to prerecord material and view it in a condensed time frame, allowing for more leisure. This is especially true in watching sporting events where time-outs, commercials, delays, half times and poor plays may be viewed on fast forward. This allows the sports fan to watch an entire football game in 30 minutes or a hockey match in 15 minutes (see Harvey & Rothe, 1985).

TAKING CONTROL

Instead of slouching in front of the screen, passing time, viewers are now participants in the creation of the television viewing experience. People are watching exactly what they want to watch. A wide variety of prerecorded material is available from which to choose, such as video cookbooks; astrology; nutrition and exercise; speech excerpts from Kennedy,

Churchill, and King; "how- to" or "self-help" programs; and so forth (see McCullaugh, 1985). In addition, as Vale (chapter 11, this volume) writes, people are also participating in the "screening of America" by videotaping their own weddings, birthdays, athletic events, and so on.

Collins (1988) provided reactions to the "active viewer" from some of the more prominent names in the video industry when he quoted Ted Turner (whose cable channels helped shape the new viewing environment) as saying, "Unquestionably, viewers have more power and control, because they have more choices." Programming innovator Fred Silverman agreed that "viewers seem to be more and more active" (p. 13). Researchers are also involved in observation of the viewing activity phenomenon. Mark Levy (1983) talked about VCRs "creating a new age for mass communication, an era in which mass media audiences have more message choices, greater involvement with the media and increased control over the timing and general experience of exposure" (p. 265).

There are only so many hours in a week, and only so many of them represent free time, and only so much free time is spent in front of a set, so how are broadcasters reacting to the newly empowered viewer? They are inventing new ways in which to capture and keep their attention. The most obvious approach is in the improvement of commercial programming. But according to Collins (1988), a variety of other innovations have also been introduced to curb what the commercial broadcasters call "destructive" viewing activities. "One requires the TV networks to present trivia questions or puzzles before the first commercial in the pod is aired and then present the answers at the end of the pod" (p. 13). "Pods" are seen as a way to curb zapping and zipping. The idea proposes that people will "stay tuned" and watch all of the commercials in between the mini-questions or presentations. A common example of a pod would be the sports-oriented, "You Make the Call." Other approaches touted by the Association of National Advertisers include the following: prize games to guarantee viewer participation during commercial breaks, such as a sweepstakes based on Social Security numbers; using a split-screen technique to run commercials while programming continues with little or no sound and action; presenting a bingo-type game using the advertised products instead of number called TV WINGO; or, grouping compatible commercials within the same pod, such as a wine commercial with a cheese commercial ("Making the call on pods," 1986).

Interactive television may be another way to recapture lost viewers, and not only with home shopping shows. The commercial, over-the-air broadcasters are trying interactive children's games, and, as in the case of the February 16, 1989 episode of "Matlock" on NBC, interactive drama where viewers were invited to phone in and vote on who was the murderer. Simply put, as audience control over leisure viewing has changed, so too has the broadcast industry as it attempts to maintain its share of that free time.

The leisure-time activity of going to the movies has also been affected by the VCR. Observers such as Seideman (1986) and Robbins (1985) claimed the film industry is suffering due to VCR usage. Seideman quoted Market Facts Inc., saying "home video is rapidly becoming the leading medium through which Americans watch feature films. At least 40% of feature-film viewing is done with VCRs" (p. 39). Robbins agreed, stating that "after the first two years of box office revenues climbing despite VCR sales, now box office revenue is down while VCR sales are up" (p. 23). Others disagree, saying "VCRs were originally feared as a threat to the American film industry, but movie house attendance in the U.S. has actually been stimulated by the VCR, in much the same fashion as Music Television (MTV) fuels the sale of rock records and videos" (Howell, 1986, p. 291). Some grist for this debate can be found in a survey conducted by the Motion Picture Association of America that discovered a 56% increase in movie attendance from 1986 to 1987 by people over 40 years old. Denby (1988) attributed this surge to the VCR-created "appetite for seeing movies where they belong, in their home, their temple and universe, the movie theatre" (p. 39).

VCRs, CHOICES, AND LIFE SATISFACTION

The relationship between television and leisure time was straightforward and obvious. World War II industrialized and urbanized America, so the work day and work week were no longer bound by the sun and seasons, but by the clock. Americans then had more time for leisure pursuits. In addition, as people left the farms to work in the employ of others, they had less need to put their earnings back into their livelihoods, freeing greater amounts of money for leisure. The newly emerged yet technologically, programmatically, economically, and organizationally mature medium of television could be afforded by those who had time in need of filling and its entrenchment was assured by advertisers of consumer goods seeking to entice people to send their newly available discretionary dollars their way, making possible greater amounts of attractive programming.

Television, then, logically became a perfect free time filler; but, in part because of people's relative lack of control over its content and scheduling, it was seen as just that. After analyzing the 1965 leisure time data, Robinson (1969) wrote,

> Perhaps few viewers consciously plan their TV evenings ahead of time by marking off those programs in *TV Guide* which look most appealing. Rather, if the weather is unaccommodating, if one can find nothing better to do, if one is looking for an hour or two of relaxation after a hectic day at work or with the children, if one is waiting for friends to call . . . what better noninvolving activity is there than an innocuous TV program that may be tuned in at any time?" (p. 218)

In his examination 10 years later he asked his respondents to rate how much they "liked or disliked participating in various obligatory and free-time activities, on a scale running from 0 (dislike a great deal) to 10 (like a great deal)." He discovered that "while the average score of 6.1 for TV viewing did fall on the positive side of the scale, it was well below the scores of almost all other free-time activities." He concluded, "These data make it hard to conclude that by increasing their viewing, the public had enhanced the subjective quality of their leisure time or their lives in general" (Robinson, 1981, p. 129).

The relationship between VCRs and leisure, however, does suggest that this subjective quality of leisure has been enhanced by the introduction of that technology and its use with home television receivers. Harvey and Rothe (1985) offered, "Not since the advent of the television set has any home electronic device began to make such a profound impact on the way Americans spend their leisure-time hours than videocassette recorders" (p. 19). Howell (1986) wrote, "Home video technology decreases the transience of television broadcasts and shifts much of the control over choice of programming and viewing time from the broadcaster to the user of the TV receiver" (p. 286).

Remembering Tinsley et al.'s (1977) linking of the amount of choice and selection within leisure time with individuals' life satisfaction, it is easy to argue that VCR is more than a free-time filler; it is free time, and therefore life-quality enhancer.

Where television helped fill free time, the VCR shifts the locus of control over much of our free time to us. This is an important function given either perspective on the amount of leisure available to Americans. If we have less time, the VCR allows us to make more efficient and personally meaningful use of what we have. If we have more, but it is harried by too many choices or too much to do, then it allows greater control. In either case, the VCR empowers viewers, affording them greater control over and choices within their free time, and possibly enhancing their life satisfaction.

Chicago *Sun Times* columnist Judy Markey made this point in a somewhat tongue-in-cheek manner. She wrote,

> When VCRs came out, we crossed over some sort of ethical edge. Because suddenly it was actually possible to be out until midnight eating sushi and still see "thirtysomething." I mean it used to be that acts (eating sushi) had consequences (missing "thirtysomething"). But suddenly what happened was that acts (eating sushi) had second acts ("thirtysomething" on tape at midnight). (1988, p. B-2)

Instead of simply filling available leisure time (as did television) or providing more attractive viewing options for today's shrinking amounts of leisure time (as was predicted for it), the VCR has put us ultimately in charge of the quantity *and* quality of a good deal of our leisure time. Moreover, there is every indication that the VCR will become even more central to our control of that free time. The technology is being improved constantly, assuring even easier and more enjoyable use and, therefore, increased options and satisfaction. Super VHS tape machines (SVHS) were introduced in 1988 by Japan Victor Corporation, utilizing 420 lines of horizontal resolution rather than the traditional 330 lines. This provides a dramatically clearer picture. Two-deck machines are now available, making possible home cassette-to-cassette dubbing and off-station taping while viewing tapes. In November 1989, ABC and machine manufacturers began experimenting with bar-code programming, invented by Panasonic in 1987. The network inserts a bar-code in its *TV Guide* ads, and viewers need only pass a reading wand over it to automatically program their VCRs to tape the specific program they want, making time-shifting even more common (Atkinson, 1989).

Enriching viewers' leisure options even further, the broadcasters, ratings organizations, and advertisers are all working to better understand audience needs, wants, and viewing patterns. Presumably programming will more closely follow viewer demands and expectations, providing us with even more attractive options and choices.

It seems clear that the VCR has already affected the ways in which Americans are spending their leisure time. But what will this mean for leisure-time patterns of the future? What will it mean if people are spending more time at home, or spending more time watching programs on the VCR with family members or friends? Will American leisure-time patterns differ from those in other countries, and to what extent might this be correlated with patterns of VCR penetration? What will become of other traditional leisure-time activities if we spend more of what little free time we have with the VCR? These and other questions need to be investigated as VCRs are found in more homes throughout the world.

REFERENCES

Atkinson, T. (1989, July 12). VCR programming: Making life easier using barcodes. *Los Angeles Times,* VI, p. 1.

Baran, S. J., McIntyre, J. S., & Meyer, T. P. (1984). *Self, symbols and society.* Reading, MA: Addison Wesley.

Collins, G. (1988, March 20). For many, a vast wasteland has become a brave new world. *New York Times,* p. 13.

DeLuca, S. M. (1980). *Television's transformation: The next 25 years.* San Diego: A. S. Barnes.

Denby, D. (1988, June 6). Fatal attraction: The VCR and the movies. *New York,* pp. 28–39.

Gibbs, N. (1989, April). How America has run out of time. *Time,* pp. 58–67.

Harvey, M., & Rothe, J. T. (1985). Video cassette recorders: Their impact on viewers and advertisers. *Journal of Advertising Research, 25,* 19–27.

Hornik, J., & Schlinger, M. J. (1981). Allocation of time to the mass media. *Journal of Consumer Research,* 7(4), 343–355.

Howell, W. J. (1986). *World broadcasting in the age of the satellite.* Norwood, NJ: Ablex.

Kim, W. Y., Baran, S. J., & Massey, K. K. (1988). Impact of the VCR on control of television viewing. *Journal of Broadcasting and Electronic Media, 32*(3), 351–358.

Levy, M. R. (1983). The time-shifting use of home video recorders. *Journal of Broadcasting, 27*(3), 263–268.

Making the call on pods. (1986, March 24). *Advertising Age,* p. 17.

Markey, J. (1988, October 21). VCR = Very complex recreation. *San Jose Mercury News,* p. B–2.

Marvin, C. (1988). *When old technologies were new.* New York: Oxford University Press.

McCullaugh, J. (1985, December 28). A day in the life of a video family. *Billboard,* pp. T8, T40.

Nayman, O. B., Atkins, C. K., & Gillette, B. (1973). The four-day workweek and media use: A glimpse of the future. *Journal of Broadcasting, 17*(3), 301–308.

Nielsen Media Research. (1988, Fall). *Nielsen newscast.* Northbrook, IL: Author.

Richard, J. (1988, November 28). Out of time. *New York Times,* p. A25.

Robbins, J. (1985, November 27). Fear about VCR threat renews. *Variety,* pp. 3, 23.

Robinson, J. P. (1969). Television and leisure time: yesterday, today and (maybe) tomorrow. *Public Opinion Quarterly, 33*(2), 210–222.

Robinson, J. P. (1981). Television and leisure time: A new scenario. *Journal of Communication, 31*(1), 120–130.

Robinson, J. P. (1989). Time's up. *American Demographics, 11*(7), 33–35.
Rojek, C. (1985). *Capitalism and leisure theory.* New York: Tavistock.
Seideman, T. (1986, March 8). Study shows VCR film viewing grows. *Billboard,* p. 39.
Tinsley, H. E. A., Barrett, T. C., & Kass, R. A. (1977). Leisure activities and need satisfaction. *Journal of Leisure Research, 9*(2), pp. 110–120.

6

VCRs and the Effects of Television: New Diversity or More of the Same?

Michael Morgan
James Shanahan
Cheryl Harris
University of Massachusetts—Amherst

A very long time ago, in a discussion of the constraints imposed by the standard, day-to-day routines of network broadcasting, Todd Gitlin (1979) noted that

> capitalism provides relief from these confines for its more favored citizens, those who can afford to buy their way out of the standardized social reality which capitalism produces. Thus, Sony and RCA now sell home video recorders, enabling consumers to tape programs they'd otherwise miss. The widely felt need to overcome assembly-line leisure time becomes the source of a new market—to sell the means for private, commoditized solutions to the time jam. (p. 255)

Gitlin wrote those words in 1979, in an era that now seems to be ancient history given the technological developments of the 1980s. Although institutional and market changes have been such that the VCR is no longer a privilege of the elite, the notion that VCRs turn television into even more of a "product" remains compelling, especially in the sense that with the VCR, the "product" becomes truly reusable and recyclable.

Thinking about media content as a product raises a host of important questions in relation to VCRs. Does the VCR introduce more or less routinization into the family's video consumption practices? To what extent do VCRs allow viewers to escape the confines of mainstream entertainment? Does the VCR amplify or fragment the cultivation of dominant conceptions of social reality? Does it imply greater or lesser family atomization? Does it break up or consolidate the viewing ritual?

Does the VCR in any way change the role of media in the processes by which family relations and interaction patterns reflect predominant structures of social organization and maintain prevailing ideologies?

Although not all of these questions are fully or easily answerable, it is clear that the VCR and some related communications technologies have obviously and dramatically changed the home media environment. In a remarkably brief time, the VCR has been transformed from an esoteric and expensive contraption to an indispensable and normal accoutrement to a television set. The family without a VCR is now the exception, especially when children are present. The U.S. Bureau of the Census (1989) reported that VCR ownership jumped from 20.8% in 1985 to 58.1% in 1988. Video rental outlets seem to have popped up on almost every street corner across the United States, from small rural areas to big city downtowns. VCRs have contributed to the drop in audience share (and revenue) among the three major broadcasting networks (Lawrence, 1989), and have profoundly altered the marketing and distribution of "theatrical" films.

To the typical television viewer, the VCR may have changed forever the way television is perceived and used. Through its time-shifting capabilities, VCRs allow viewers to watch broadcast and cable programming whenever and as often as they like. They may thus feel a new sense of power and control over their viewing fare, derived from the ability to freeze a frame, review a scene, zip through commercials (or zap them entirely) and so on. Moreover, through renting prerecorded cassettes, viewers may now believe they have an unprecedented range of choice in what they select to watch. These changes in the traditional meaning of "watching TV" raise questions about the continued legitimacy of conceiving of media effects only in terms of television. In this chapter, we explore these issues in the context of cultivation theory.

CULTIVATION

The theory of cultivation, developed by Gerbner and his colleagues (Gerbner & Gross, 1976; Gerbner, Gross, Morgan, & Signorielli, 1980, 1986) is based on the assumption that television defines the mainstream of American culture. The cultivation perspective assumes that, compared to other media, television is used relatively nonselectively, and that television presents a relatively restricted and repetitive set of messages, images, and values about life and society. Thus, unlike many other approaches to media effects, cultivation analysis is not concerned with specific programs, episodes, or genres, and instead assumes that overall exposure to television is the most critical indicator of potential effects.

Usually based on survey research methods, cultivation analysis attempts to determine the extent to which people who watch greater amounts of television ("heavy viewers") hold different conceptions of social reality from those who watch less, other things held constant. The basic hypothesis of cultivation research is that heavy viewers will be more likely to perceive the real world in ways that reflect the most stable and recurrent patterns of portrayals in the "television world," as revealed through annual, systematic analyses of television content. Amount of television viewing has been found to make an independent contribution to people's beliefs, assumptions, and values in a broad range of substantive areas, including images of violence, mistrust, sex-role and age-role stereotypes, the family, health, religion, science, political orientations, and many other issues (Morgan, 1989; Signorielli & Morgan, 1990).

At first glance, the VCR may appear to strongly challenge or even to negate some assumptions of cultivation theory. Armed with a VCR, now the viewer at least *can* be more selective than ever. Instead of being limited to whatever happens to be on the air, viewers can pick and choose what they want to record or rent from a vast range of alternatives. Yet, this scenario assumes that the specific *content* that VCR users (especially heavy VCR users, and especially those who are heavy television viewers in general) will attend to indeed presents *alternative* world views, values, and stereotypes from most network-type programs.

Given the tight links among the various industries involved in the production and distribution of media content, and the fact that they are all trying to attract the same overlapping, heterogeneous audiences, it is likely that the most popular program materials will tend to present consistent and complementary messages. Moreover, industry programming practices are geared to reproduce what has already proven to be profitable (Gitlin, 1983). A caveat: We know of no empirical content analyses of what VCR owners watch that would support or refute this contention. But what is most popular, by definition, tends to reflect—and cultivate—dominant cultural ideologies. Certainly, the VCR *allows* selective viewers to seek out specialized, often "fringe" material (Dobrow, 1990); but for average to heavy viewers, most of the time, the VCR is likely to be used to consume "more of the same."

STORYTELLING, MYTHS, AND MEDIA

If VCRs allow viewers (especially heavy viewers) some greater convenience in when and how often they watch television and provides them with the means to watch a greater amount of similar material, then the

VCR may be more likely to extend and augment cultivation than to fragment the mainstream. This is because the theory of cultivation is really much more a theory of the effects of storytelling than it is a theory of the effects of television. Television is studied for this purpose, not as a technology per se, but as a central institution of cultural production, because it is the most pervasive source of standardized, market-driven, centrally produced cultural stories in this society.

That the content of stories should have an effect on what people believe about their world is not a new concept. Indeed, literary critics, governments, educators, religious leaders, and other guardians of public morality have always dealt with issues similar to those we have been dealing with in this era with regard to television. If we accept the notion that the academic study of television is a revised approach both to mythology and storytelling, then cultivation analysis is a way of doing literary and mythographical criticism in the modern (or post-modern) era. In this sense, cultivation theory is merely one frame for assessing the impact of how we entertain and inform ourselves (a distinction television has blurred; Meyrowitz, 1985). Because the approach is well-defined theoretically and supported by a concrete methodological system, there may be a tendency to isolate it from other, "qualitatively oriented" systems of thought. But cultivation theory owes as much to the traditions of literary criticism, religious study, and mythography as it does to quantitative social science and abstracted empiricism.

For instance, Roland Barthes, at one time one of the leaders of the French avant-garde in cultural criticism, often presented arguments highly congruent to cultivation theory when speaking about the nature of myth in modern French society (Barthes, 1972). Barthes, seeing myth as a conjointly enacted product of a culture and its media, argued that myth is a form of depoliticized speech, and that the form that speech takes has important effects on the way the community receives the speech. Most essentially, he argued that the incessant, continual use of certain forms of speech, and an unwillingness by the speakers to reveal attributions about their motives, results in the creation of new myths, myths that conceal the disjuncture between "Nature and History." In essence, he believed that mass culture cultivates its own acceptance.

The cultivation perspective argues that people will absorb notions about reality that are not necessarily "true," as long as these notions are couched in the "realistic" style of conventional Western narrative. Thus, cultivation studies a mode of speaking, which we call *representational realism.* Representational realism is the preferred style of discourse on television, and probably the dominant modality for effecting cultivation. Because television is incorporated into standardized American modes of generating reality, those who do "speak" via television wield a privileged sort of influence.

Seen from this kind of cultural perspective, cultivation analysis is neither a sterile nor inflexible set of methodological procedures, only suitable perhaps for discussing the social impact of a certain rigidly defined type of technology. Rather, cultivation is as much a critical theory (in the current sense of that term) as any linguistic or political approach to the social construction of power and reality.

FORM, CONTENT, AND VCRs

All of this, in the context of the VCR, introduces some new concerns for cultivation theory. Certainly, some technical, formal aspects of the television medium, along with the economic and institutional features of its organization, have contributed heavily to its consequences. The very fact of *broad*casting is one such factor in television's effects, along with its repetitive nature. For certain analysts, formal qualities have remained the preeminent focus in the study of the effects of technology (e.g., Kittler, 1987; Olson, 1988). An overemphasis on technological determinism, however, could lead to the conflation of television with only its technical aspects, and in turn to the view that any significant changes in the technology used to deliver television would necessitate significant changes in its effect. A different analysis, less focused on form and more concerned with content, would at least be willing to investigate the ratio of the actual relationship between form and content in shaping effects.

These are some of the types of questions that confront cultivation theory as various developments in the entertainment industry appear to be changing the ways in which stories are told in America (although, as we argue, without much change in the stories themselves). The introduction of various home-entertainment technologies, especially the VCR, again has the potential to allow more diversity and selectivity in the choices Americans make in their entertainment decisions. To some, the drop in network audience share brought about by the VCR and various other "delivery" systems might mean the "death" of cultivation, because of new and diverse viewing options. To others, the term *delivery* system is especially telling because it implies that changes in delivery of programming do not necessitate changes in content and effect. To twist a well-worn phrase, this may be a case of old wine in new bottles.

The idea of "expanded delivery," with less focus on the technical aspects of transmission and reception, suggests no lessened relevance for cultivation theory in the future, especially because expanded delivery systems may actually *strengthen* cultivation, by increasing the time we spend absorbing mass-produced myths. For example, VCR owners tend to watch more television overall (Murray & White, 1987). Clearly, ownership of both cable and VCRs allows viewers not only to watch new kinds of specialized programming more often (especially movies) but also reflects a greater commitment to video entertainment in any format, which explains the association between over-the-air viewing and cable and VCR ownership. Overall, of course, mass-produced myths are not limited to network television broadcasting, so any ownership of new media technologies is likely to increase exposure to such material. Indeed, at least one study has found that cultivation patterns are notably stronger among those who also have access to cable television (Morgan & Rothschild, 1983).

One key aspect of cultivation theory is that the most important media effects are those that accrue, cumulatively, over time. Although individual programs may seem to effect change at times, fundamental cultural effects are best measured at the aggregate level, reflecting the stabilization of cultivated norms. The expansion of delivery systems could actually contribute to such a stabilization, but if and only if the content matter remained essentially similar. Otherwise, competition between differing content patterns might attenuate cultivation, by cultivating distinct and pluralistic publics based on a variety of special interests, as is the case with magazines and other print media.

"NEW" TECHNOLOGIES?

It is odd that there is a general failure to recognize the fairly obvious historical precedent to these arguments. Since the inception of electronically mass-delivered entertainment programs in this country, there has been really only one notable shift in the mode of program delivery: the changeover from radio to television. Far from destroying either radio or film, television strengthened the maintenance of a system of norms in the minds of American audiences as it adopted the program forms and genres developed in those earlier media. And yet, in hindsight, we can see that the emergence of television is precisely the type of technological development about which many are currently concerned, if indeed it is not a stronger example of such a case. If such a radical transformation in style failed to produce significant change in cultivated norms (or even strengthened those norms, as seems more likely) why does a similar development—the VCR—provoke concern today?

Undoubtedly, some of the hullabaloo surrounding the VCR must be attributed to the entertainment and information industry itself; although the industry has been concerned with the impact of the VCR on established media markets, it has also fostered the idea that America is entering a utopian "information age," in which the commodities and services that the industry sells will play a key role. The economic motives for expanding delivery systems are patently clear to most analysts of the media industries: While delivery modes are expanding, ownership "modes" are contracting. The saturation of the market for television (both as a technology and as a mode of speech) is the immediate precursor to the development of apparently "alternative" technologies (which most have assumed will somehow automatically bring "alternative" modes of speech).

Politically and financially, information industries are motivated to encourage the idea that expanded technologies will contribute to expanded opportunities for self-expression, a kind of co-optation of the "marketplace of ideas" approach. However, the FCC, which has been clearly committed to the marketplace as a shaper of ideas, refuses even the most basic restrictions on the growingly more condensed pool of licensed "speakers" (Ferrall, 1989). For instance, cable deregulation has not led to a proliferation of ways to speak and channels for self-expression in this country; the growing view in Congress and elsewhere is that the results have been increased rates and decreased public service. However, various modes of consuming old sitcoms and wrestling have become more available.

It is in this atmosphere that we encounter what used to be called *new technologies.* We can see more easily now (as Gitlin saw in 1979) the extent to which this term is the creation of financial analysts, administrative researchers, and marketers interested in using the new technologies to increase profits in the entertainment industry. This atmosphere fostered a tendency to conceive of freedom in terms of channels and bits (Dizard, 1982), rather than the social content of those channels or who owns them. But this sort of analysis failed to appreciate the stability in programming values, because attention was, by the very definition of the approach, diverted from content matters.

Partially because of this situation, most of the literature on VCRs emphasizes structural and economic concerns over the social. There are many descriptive analyses of how VCRs are used (e.g., Levy, 1980; Rubin & Bantz, 1987). The general consensus has been that VCRs allow greater involvement for the viewer, both in terms of restructuring the times programs will be viewed, as well as bypassing unwanted material. More recently, some studies have addressed social aspects of VCR use, including co-viewing (Gunter & Levy, 1987), its relation to other viewing behaviors and rules (Brown, Bauman, Lentz, & Koch, 1987), and possibilities of VCR use for cultivation (Dobrow, 1990). With research on the VCR beginning to gain momentum, we may expect less concentration on technical and procedural questions and more focus on social issues. To date, however (with some notable exceptions, such as Roe, 1987), that has not largely been the case.

Yet, it should be clear that economic concerns influence programming values. Although the changes in technology may herald superficial changes in the *dramatis personae* of the media scene, it is unlikely that these technologies will reverse the increasing trend toward concentration in the media industries. Thus, there is no corresponding change in programming values, or even in the particular style in which "television" is spoken. As Gerbner (1990) said "Always touted as the dawning of new freedoms, new technologies typically penetrate new markets and eventually concentrate money, power, and choices. To that extent, they may intensify rather than dilute the central thrust of the cultivation process."

If these assumptions are true, then the logical way to deal with the effect of new technologies, from the cultivation perspective, is to investigate the extent to which they maintain the stability of the "old" system of programming values, a system that is well-understood from an enormous body of mass communication research. Various relationships previously established in cultivation research, such as the maintenance of a host of stereotypical values by TV (sex roles, occupations, attitudes toward violence, etc.), should be sustained or augmented by the introduction of alternate ways to deliver the same modes of speech. This should hold for VCRs, cable, and any other "new" technology that we may encounter in the coming years.

RESEARCH QUESTIONS, PROCEDURES, AND MEASURES

Our opportunity to put some of these ideas to empirical tests comes from a longitudinal study we have been conducting on the acquisition, usage, and impacts of new technologies in the home (Alexander, Morgan, Shanahan, & Harris, 1989; Morgan, Alexander, Shanahan, & Harris, 1990). Based on data from this broader study, we have explored relationships between television viewing, VCR use, and cultivation, in order to test the hypothesis that cultivation will be maintained or increased when VCR fare is added to the media diet.

In particular, we present some preliminary findings from this work here, which address several research questions: How do ownership and amount of use of the VCR intervene in the cultivation process? Do VCRs maintain and strengthen cultivation, or do they decrease the contributions of traditional television's messages to viewers' conceptions? How does the social context of VCR usage enhance, diminish, or otherwise mediate cultivation? Do particular uses of the VCR, such as time-shifting and renting movies, mediate cultivation in particular ways? Most of all, is it meaningful to continue to conceive of various popular media as having distinct and separate influences, or are different electronic entertainment technologies functionally equivalent in their cultural and cultivating consequences?

The data from which we are drawing to explore these issues come from a longitudinal study of adolescents. In 1985, surveys were administered to 910 students (Grades 7 through 12) in a southeastern New England town. The survey concentrated on the uses and impacts of new media technologies in the family context. In 1988, we returned to this site and gathered new data with a special concentration on VCR use, with a sample size of 642 and a repeat sample of 206 adolescents. In both waves, we measured overall television exposure for several time periods: on "typical school days" after school and before dinner, after dinner and before bed, on Saturdays, and on Sundays.

Standard cultivation measures about violence, interpersonal mistrust, sexism, and educational expectations were also included in both surveys. The violence question asked respondents whether they thought their chances of being involved in violence were 1 in 10 or 1 in 100. The "mean world syndrome" of mistrust was measured in terms of whether respondents thought most people can be trusted or that you "can't be too careful" in dealing with people. Sexism was measured in terms of the extent to which respondents agreed or disagreed with the notion that "By nature, women are happiest raising children and caring for the home" (cf. Morgan, 1982, 1987). Additionally, demographic information was obtained, including educational and occupational levels in the family and adolescents' expectations. We recognize that adolescent attitudes may be strongly influenced by factors such as peer pressure. However, we were primarily interested in describing associations between attitudes and viewing in relative, not absolute, terms. Thus, the validity of adolescents' responses is not radically threatened.

For the second wave, the instrument was tailored to address VCR concerns, partly because of the obvious need for longitudinal data regarding VCR use. In addition to the categories described here, a wide range of questions was introduced to measure particular aspects of VCR use in the second wave. These included respondent use ("almost every day" to "hardly ever," a 5-point scale), VCR co-viewing patterns, types of uses of the VCR endorsed by the respondent (i.e., cable movies taped, late night TV, etc.), decision making and conflicts about the VCR, and rules about the VCR within the family.

SOME FINDINGS

Drawing from both cross-sectional samples, we found some dramatic, and not unexpected, changes in VCR "penetration" over the 3 years between our data collections. In 1985, VCRs were available to less than one third (29.9%) of the adolescents in our sample, but in 1988, 89.4% of the respondents' families owned VCRs. These figures are somewhat higher than national figures for families with children (Lin & Atkin, 1989) but they are also slightly inflated due to the presence of siblings in the data set.

In 1985, the novelty of the technology was reflected in the fact that very few (15.7%) of those with VCRs had owned them for more than 2 years, whereas a slight majority (56%) had owned them for less than 1 year. By 1988, the VCR was no longer new and unfamiliar. Novelty effects had a chance to wear off and usage patterns were likely to have become stable by 1988; fully half of those with a VCR had owned it for more than 2 years, and hardly any (11.2%) had owned it for 1 year or less. Almost 79% of the adolescents reported that their family used the VCR at least a few times per week in 1985, but this figure softened somewhat, to more than two thirds (68.4%), in 1988, perhaps reflecting a stabilization of use over time.

Renting theatrical movies was the most common use reported in both years, although there was a very large increase in time-shifting. The adolescents also reported a variety of new things they were able to watch as a result of owning a VCR, such as late night TV, soap operas, and rock videos. Hardly any (3.9%) said they used the VCR to watch "no different shows," an indirect indication that adolescents perceive the VCR as expanding their viewing options. The VCR thus allows many adolescents to view a greater amount of network-type programs while it gives them a greater *sense* of diversity and choice.

Not surprisingly, the heaviest VCR users are also the heaviest television viewers (in the 1988 survey, the simple correlation of amount of television viewing with amount of VCR use is .31, $p < .001$). Examining the data over time (using the panel data), we found that amount of television viewing in 1985 had a significant impact on VCR ownership in 1988 (controlling for gender, grade, parental education, and VCR ownership in the first wave), but that VCR ownership did not affect television viewing levels 3 years later. This suggests that television viewing is a "driving force" behind VCR ownership, and that VCR patterns are likely to grow out of television usage patterns (because having a VCR does not change amount of television exposure over time).

In other words, over a 3-year time span, the acquisition of a VCR did not seem to affect how much time adolescents spend watching television. This can also be seen in terms of the viewing changes over time of groups defined by change and stability in VCR ownership, as seen in Fig. 6.1. In 1985, those who either owned a VCR or would own one by 1988 tended to watch more television, but by 1988 these differences disappeared. (The mean viewing figures are calculated from composite measures of self-reported hours of viewing after school, after dinner, and on weekends; they should be interpreted in relative terms, and should not be seen as "accurate" reports of actual viewing hours.) The steep declines in amount of viewing over time for all three groups reflect the well-known finding that television viewing drops sharply during adolescence (Comstock, Chaffee, Katzman, McCombs, & Roberts, 1978); VCR acquisition does not appear to significantly slow down the decrease in viewing that usually accompanies later adolescence.

Thus, heavy television viewing leads to VCR acquisition and usage. Moreover, heavy television viewers use the VCR more often for all purposes measured (renting tapes, viewing owned movies, and watching recorded TV programs) and in all social contexts (with parents, siblings, peers, *and* alone). All of these relationships are strong and significant.

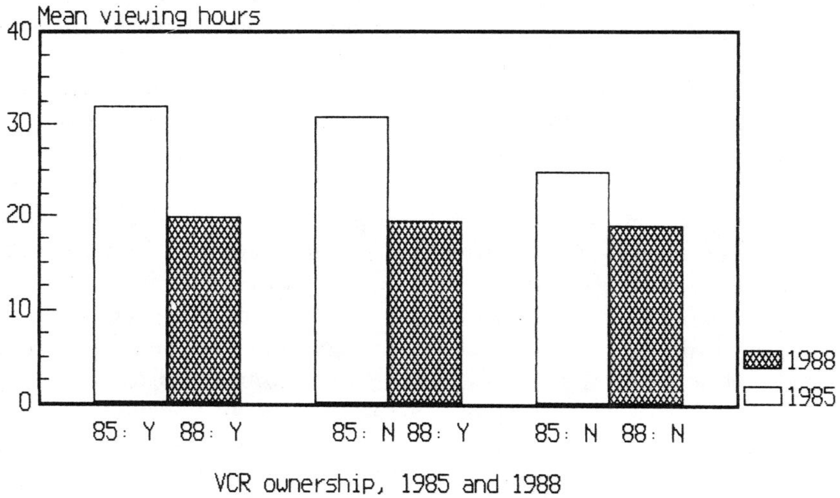

FIG. 6.1. Mean weekly TV viewing, over time, by VCR ownership patterns (N = 206).

Also, heavy viewers are more likely to use the VCR to watch more late night television programs, more rock music videos, and more of "my favorite shows."

VCRs AND CULTIVATION

Given these general findings, we explored the implications of the VCR for cultivation analysis in two ways. First, we used amount of VCR usage as an independent variable, to see how the frequency of VCR usage is related to conceptions of social reality. Second, we examined the relationships between amount of VCR usage and attitudes separately for light and heavy television viewers, following the findings of Dobrow (1990) that light and heavy television viewers differ in the extent to which they use the VCR to diversify or extend (respectively) their viewing habits. For both of these purposes, we focused on sex-role stereotypes, the "mean world syndrome" of perceptions of violence and mistrust, and educational expectations.

This analysis used the two synchronous samples. Overall, there is a small but significant positive association between amount of VCR usage and sex-role stereotypes (r = .07, p = .05), and a moderate negative association between amount of VCR usage and educational expectations (r = -.19), $p < .001$). "Mean world" measures are unrelated to overall amount of VCR usage, for both males and females. The association of VCR usage with sexism applies mainly to males, but the association with educational expectations holds strongly for both males and females. In general, controls for gender, grade in school, and parental education reduce the associations between overall amount of television viewing and these attitudes. With VCR viewing by the respondent as the independent measure, these control variables have much less impact.

The cultivation of sex-role stereotypes is either strengthened (r = .12, $p < .05$ in 1985) or maintained (r = .09, $p < .05$ in 1988) among respondents who owned a VCR. However, in families without a VCR, amount of television viewing is not significantly related to sexism in either year. The correlations, although extremely modest, are consistent between waves in terms of both direction and magnitude, lending greater credence to the notion that owning a VCR can augment cultivation.

When compared within light and heavy overall television viewing subgroups, these patterns become far more complex, as shown in Table 6.1. In general, the associations between amount of VCR usage and these dependent variables are stronger among those who watch more

TABLE 6.1
Scores on Dependent Attitudinal Variables by Amount of Television Viewing and VCR Usage (1988 Data; N = 550)

	Lo Overall TV		*Hi Overall TV*	
Overall VCR Usage:	*Lo*	*Hi*	*Lo*	*Hi*
(N)	*(177)*	*(98)*	*(110)*	*(165)*
Sexism	.15	.22	.15	.26
Trust	.71	.72	.68	.74
Violence	.76	.62	.59	.71
Educational Aspirations	3.12	2.84	2.93	2.81

Notes:
Overall TV Viewing: Composite weekly hours of viewing, dichotimized at the median
Overall VCR usage: Hi = "Almost every day" and "A few times a week"; Lo = "About once a week," "About once a month," and "Hardly ever"
Sexism: 1 = Strongly Agree or Agree that "By nature, women are happiest raising children and caring for the home"; 0 = Disagree or Strongly Disagree
Trust: 1 = "You can't be too careful in dealing with people"; 0 = "Most people can be trusted"
Violence: 1 = Estimate of number of people involved in violence each week is "1 in 10"; 0 = "1 in 100"
Educational Aspirations: 4-point scale, where 1 = "I won't finish high school" and 4 = "I will go to graduate school after college"

television, except for educational expectations; how much schooling these students expect to attain is negatively related to amount of VCR usage for both light and heavy viewers. But those who use the VCR more often are more sexist, more mistrustful, and more likely to overestimate the number of people involved in violence either *mainly* or *only* if they are also heavy viewers of television in general.

By the same token (again, except for educational expectations, which shows an across-the-board association), these data also show that the potential of television viewing to cultivate these attitudes, although generally weak in this sample, can be systematically enhanced by various VCR usage patterns. The associations between television viewing and these attitudes (especially sexism) tend to be markedly stronger among those who own VCRs, those who have owned them longer, those who use them more often, those who do not use them with parents or siblings, and those who do not have family rules about VCR usage.

The overall implication is that ownership and use of the VCR signifies greater commitment to video as a form of entertainment, and greater exposure to (and absorption of) consistent messages about life and society. In both waves, despite different rates of penetration of the VCR and different use rates at the different waves, the VCR points either to strengthening or maintenance of cultivation.

Finally, the specific uses to which the VCR is put (e.g., time-shifting vs. film rentals) may mediate cultivation in different ways. A priori, we would expect that cultivation would be enhanced for families who use the VCR to supplement television (time-shifting), while the renting of films and other uses might pose a threat to such cultivation. Analysis of variance shows that there are indeed some significant effects for renting versus time-shifting.

As expected, the difference in sexist attitudes is greater for those who use the VCR heavily for off-air viewing than for renting tapes ($p = .03$), but we also found significant interactions between heavy time-shifting and renting uses, suggesting that overall heavy VCR use is tied to cultivation, and probably to overall viewing as well. The use of the VCR for renting indicates less cultivation in some cases, but there is substantial incidence of a phenomenon that we could simply call "heavy overall VCR use." This phenomenon will also produce cultivation, although apparently less for those who fail to use the VCR for any TV taping purposes.

These findings suggest that, although there are many new forms of watching television and some new genres, it is too soon to conclude that there are also new programming values and ideologies incorporated into those new ways of watching. It appears that the VCR strengthens the effects of television; it does not mitigate those effects. At the same time, VCRs cultivate "television-type" conceptions mainly among those who are heavy television viewers. Altogether, these data shed some light on ongoing debates about the supposed ability of new technologies to open up a new and diverse marketplace of ideas.

CONCLUSION

This chapter argues that the VCR provides the entertainment industry with some new, and potentially effective, ways to disseminate material that is essentially similar to what was being disseminated before. Although it is true that there is little data about the content of what is being seen on VCRs, we assume that the message systems of the VCR and television are similar. In some ways, VCR programming may be even more explicit with regard to violence and sex than over-the-air television can be. Having demonstrated some similarities in terms of the effects of differential viewing, it would now also be logical to assume a functional equivalence between TV and VCR programming values. Further research should seek to demonstrate this.

Beyond this presumed functional equivalence, there are interesting and important sociological and policy implications from these data. It has traditionally been argued that later adolescence is a time of reduced TV viewing; this reduction might also be manifested in some temporary reductions in terms of cultivated effects. That is certainly the case between television viewing and sexism in the 1988 wave of this sample, where older adolescents were somewhat less susceptible to the effects of overall viewing with respect to sexism. But the addition of new technologies changes that picture. This points the way toward a new conception of overall media use as an independent variable in media effects research.

One common criticism of cultivation has been the supposed one-dimensionality of the simple measure of exposure to television. These data demonstrate that the similarities between television and other like technologies are more pronounced than their differences. In some cases, it may even be that new technologies are absorbing old roles in the maintenance of media effects. Even if it is the case that simple TV viewing has a lesser effect on adolescents, other media such as the VCR appear to "pick up the slack." This suggests not only that the entertainment industry is successful at finding new ways to attract even "obstinate" audiences to their messages, but also that traditional messages can be transmitted in nontraditional ways, yet with decidedly traditional results.

These data also illuminate the role that cultivation analysis may play in the continuing effort to provide a critical analysis of the effects of new media technologies. Although the approach rests on quantitative schemes of analysis, it should be clear that a "critical" interpretation is readily available for these data. Perhaps it is not too difficult to see how even Barthes might have appreciated the advantages of such an analysis. In this context, we have found it useful to divorce *cultivation* from *television* specifically, despite the common association between the two terms. Cultivation then becomes a statistical and analytical account of the "teaching potential" of modern industrial storytelling, regardless of the medium used for delivery.

REFERENCES

Alexander, A., Morgan, M., Shanahan, J., & Harris, C. (1989). *Videocassette recorders and the family.* Report to the National Association of Broadcasters, Amherst, MA.

Barthes, R. (1972). *Mythologies.* New York: Hill & Wang.

Brown, J. D., Bauman, K. E., Lentz, G. M., & Koch, G. G. (1987, May). *Young adolescents' use of radio and television in the 1980s.* Paper presented at the Annual Meeting of the International Communication Association, Montreal.

Comstock, G., Chaffee, S., Katzman, N., McCombs, M., & Roberts, D. (1978). *Television and human behavior.* New York: Columbia University Press.

Dizard, W. P. (1982). *The coming information age.* New York: Longman.

Dobrow, J. (1990). Patterns of viewing and VCR use: Implications for cultivation analysis. In N. Signorielli & M. Morgan (Eds.), *Cultivation analysis: New directions in media effects research* (pp. 71–83). Newbury Park, CA: Sage.

Ferrall, V. (1989). The impact of television deregulation on private and public interests. *Journal of Communication, 39*(1), 8–38.

Gerbner, G. (1990). Advancing on the path of righteousness (maybe). In N. Signorielli & M. Morgan (Eds.), *Cultivation analysis: New directions in media effects research* (pp. 249–262). Newbury Park, CA: Sage.

Gerbner, G., & Gross, L. (1976). Living with television: The violence profile. *Journal of Communication, 26*(2), 173–199.

Gerbner, G., Gross, L., Morgan, M., & Signorielli, N. (1980). The "mainstreaming" of America: Violence profile no. 11. *Journal of Communication, 30*(3), 10–29.

Gerbner, G., Gross, L., Morgan, M., & Signorielli, N. (1986). Living with television: The dynamics of the cultivation process. In J. Bryant & D. Zillmann (Eds.), *Perspectives on media effects* (pp. 17–40). Hillsdale, NJ: Lawrence Erlbaum Associates.

Gitlin, T. (1979). Prime time ideology: The hegemonic process in television entertainment. *Social Problems, 26*(3), 251–266.

Gitlin, T. (1983). *Inside prime time.* New York: Pantheon.

Gunter, B., & Levy, M. (1987). Social contexts of video use. *American Behavioral Scientist, 30,* 486–494.

Kittler, F. (1987). Gramophone, film, typewriter. *October, 41,* 101–118.

Lawrence, R. (1989). Television: The battle for attention. *Marketing and Media Decisions, 24*(2), 80–82.

Lin, C., & Atkin, D. (1989). Parental mediation and rulemaking for adolescent use of television and VCRs. *Journal of Broadcasting and Electronic Media, 33*(1), 53–67.

Levy, M. (1980). Home video recorders: A user survey. *Journal of Communication, 30*(4), 23–27.

Meyrowitz, J. (1985). *No sense of place: The impact of electronic media on social behavior.* New York: Oxford University Press.

Morgan, M. (1982). Television and adolescent's sex-role stereotypes: A longitudinal study. *Journal of Personality and Social Psychology, 43,* 947–955.

Morgan, M. (1987). Television, sex-role attitudes, and sex-role behavior. *Journal of Early Adolescence, 7,* 269–282.

Morgan, M. (1989). Cultivation analysis. In E. Barnouw (Ed.), *The international encyclopedia of communications* (Vol. I, pp. 430–433). New York: Oxford University Press.

Morgan, M., Alexander, A., Shanahan, J., & Harris, C. (1990). Adolescents, VCRs, and the family environment. *Communication Research, 17*(1), 83–106.

Morgan, M., & Rothschild, N. (1983). Impact of the new television technology: Cable TV, peers, and sex-role cultivation in the electronic environment. *Youth and Society, 15,* 33–50.

Murray, M., & White, S. (1987). VCR owners' use of pay cable services. *Journalism Quarterly, 64*(1), 193–195.

Olson, D. (1988). Mind and media: The epistemic functions of literacy. *Journal of Communication, 38*(3), 27–36.

Roe, K. (1987, May). *Adolescents' VCR use: How and why.* Paper presented at the annual meeting of the International Communication Association, Montreal.

Rubin, A., & Bantz, C. (1987). Utility of videocassette recorders. *American Behavioral Scientist, 30,* 471–485.

Signorielli, N., & Morgan, M. (Eds.). (1990). *Cultivation analysis: New directions in media effects research.* Newbury Park, CA: Sage.

U.S. Bureau of the Census. (1989). *Statistical abstract of the United States: 1989* (109th ed.). Washington, DC: Author.

7

Context, Social Class, and VCRs: A World Comparison

Joseph D. Straubhaar
Michigan State University

This chapter looks at the cultural uses and impact of VCRs in two related ways. The primary effort is to examine the ways in which social class may be related to the economic and cultural choices involved in VCR use and VCR impacts. Social class is shown to be strongly related to VCR acquisition, use, and effects in a number of societies. However, social class effects are seen as taking place within certain contexts. One context is the general system of social and economic stratification, which in large part depends on patterns of economic development. The other context that seems most critical is the nature of the television system, both its structures and content. To examine these issues, several cases in Latin America are examined and compared, along with some comparisons to other regions.

The general system of stratification structures social classes, including both their access to and, arguably, many of their interests in television and VCRs. That richer people have better access to VCRs is clear, but sometimes poorer people of one nation will acquire VCRs, whereas richer people of another nation will not; that difference will probably depend on what is offered by the TV systems of the two countries. Generally, however, I argue that upper, middle, and lower classes tend to have somewhat distinct cultural interests, particularly where class stratification is accentuated. Overall, I suggest that elites tend to be more internationalized, which is reflected in their use of VCRs to view more internationalized programming.

Elites and upwardly mobile middle classes might get VCRs for purposes of conspicuous consumption, but lower middle and lower classes will probably require a larger marginal utility for using their lower incomes to get and use a VCR. Cases can be found of poor populations getting VCRs but usually because of a manifest unhappiness with what is on broadcast television.

The main method employed by this chapter is comparative: It compares several societies with different kinds of stratification and different TV system contexts, primarily but not exclusively in Latin America. Ultimately, the goal is to examine indications of how individuals have acquired, used, and been affected by VCRs, within classes, and within larger social contexts. Where available, survey and ethnographic data are used to summarize actual individual behavior.

THE INTERNATIONAL CONTEXT

This chapter cannot definitively discuss the complexities of international relations of dependence and interdependence, but it recognized these relations as an essential background or context for understanding social stratification at the national level and below. Modernization theorists originally did not think it necessary to dwell on international relations to understand different societies (Lerner, 1958; Rogers, 1976), whereas some dependency theorists implied that knowing international relations of dependency would tell virtually all that was important about the structures of a society (Chilcote, 1984).

Clearly, international ties of dependency affect both social stratification and television systems, but both social structures and television institutions also have certain varying levels of national autonomy. Major types of international effects include transfers of general economic models, notably levels of income distribution and modes of commercialization of media.

Income distribution has been empirically demonstrated to be relevant to VCR diffusion. Both overall income (GNP) per capita and relatively more even income distribution were significantly correlated with VCR penetration in a multi-country study by Straubhaar and Lin (1988). Both absolute income levels and relative income distribution are also highly related to, perhaps almost functions of, a nation's place in the world economic system. Indeed, the World Bank and most others still tend to rank countries by GNP per capita, despite decades of attempts to find better measures of overall development that might reflect overall quality of life, such as the Physical Quality of Life Index (World Bank, 1989).

Critics, such as dependency theorists, essentially agree that life benefits are determined by a nation's role in the world capitalist economy (Souki de Oliveira, 1989; Wallerstein, 1977). In contrast to traditional economic development theorists, dependency theorists are concerned that nations' relative incomes are almost locked in, that the nature of the international system affords little mobility. For them, the economic phenomenon most relevant to new media/consumption items like the VCR in most of the world is the idea of the dualistic economy. A dualistic economy is one in which a small elite concentrates the national income in its own hands so that its members may consume at more or less the level of the Western industrial societies, while much of the society is impoverished (dos Santos, 1978; Souki de Oliveira, 1989). In a dualistic economy, one might expect to find VCRs only among the very rich.

A current and relevant argument is whether such dualism is increasing or decreasing. That can also be seen as whether relative income distribution is becoming more even or more skewed over time. In line with the empirical evidence correlating relatively even income distribution with VCR penetration noted by Straubhaar and Lin (1988), one would expect a wider (or more even) distribution of income to allow more people to buy VCRs, if they wanted them.

At another level, the international economy tends to have effects on nations' television systems that also affect VCR use. In essence, there is a gradual pressure on many nations to commercialize media, particularly television. Groups adding pressure toward commercialization include transnational and national corporations that wish to advertise products, international and national advertising agencies, film, or other cultural industries that emphasize entertainment. This is true even in nations that had long resisted commercialization, such as India (Mody, 1989).

International commercialization pressure is thought by some (Souki de Oliveira, 1989; Straubhaar, 1986) to increase economic dualism and to affect income distribution. Increased exposure by national elites and middle classes to an internationalized style of consumption and to specific products common to the industrialized nations has long been thought to bring rising expectations of greater individual benefit or participation from economic growth, in particular, more income and consumption goods. What poses a threat to improvements in income distribution is that elites may well be motivated by specific commercial propaganda and the general commercialization of culture to further skew income to permit greater consumption by themselves.

Because commercialization affects TV content, it will affect demand for and use of VCRs by various social classes. It does so in a complex manner, however. We argue that in some cases, such as Brazil, commercialization of television has actually created a TV industry that continues to maintain audience loyalty despite the availability of VCRs. This is because the entertainment created by such industries, although highly commercialized, is also highly attractive to the audience and greatly lowers the "marginal utility" of adding more entertainment via VCRs for all but the upper and upper middle classes who have enough disposable income to make questions of marginal utility less important.

Noncommercial television systems can also affect VCR use. Boyd, Straubhaar, and Lent (1989) noted several Third World situations where efforts to use broadcast television for development support programming have led large parts of the audience which are more interested in entertainment to use VCRs to obtain it.

NATIONAL SOCIETY CONTEXT

Income and income distribution are clearly affected by national actors and decisions as well as by the actors in the world economy. Indeed, as Sarti (1985) noted, no one had to teach Brazilian businessmen how to exploit people, although they may well have learned a great deal from foreign companies and agencies. A major strength of Cardoso and Faletto (1979), Evans (1979), Salinas and Paldan (1976), and others associated with dependent development theories is that they try to take account of national industry and the nation state as forces that do retain or acquire power and that may reinforce or compete with transnational corporations. Similarly, the historical development and present state of national class relations is considered as important as international relationships of dependency.

National government policy can particularly play a role in accentuating or diminishing VCR use. Probably the most critical government role relevant to VCRs comes in the structuring and control of television systems, which is discussed later. However, governments also affect VCRs directly by import restrictions on VCR machines and/or videotapes, tariffs, taxes, restrictions on tape rentals, and so on. Many governments try to restrict VCRs and videotapes for political reasons, because as unrestricted potential sources of information they do in fact pose a threat to governments, such as the USSR (Boyd, 1988) or Marcos' Philippines (Lent, 1985), which try to control information (Ganley & Ganley, 1988). Indirect political reasons, such as the Malaysian government trying to restrict ethnic video content to maintain majority control, also exist. Other governments try to restrict VCRs primarily for economic reasons, because they represent for most countries an imported luxury item. Such governments often place heavy tariffs on VCR imports, which frequently leads to smuggling, as in a number of cases in Latin America (Mattelart & Schmucler, 1985). More often than not the effect of both political and economic restrictions is simply to make VCRs more expensive and restrict access to them to the elite of the society.

TELEVISION SYSTEM CONTEXT

To understand social class-based use of VCRs, one must study and understand the television system itself. Previous research based on case studies, particularly on the USSR (Boyd, 1988) and on the Third World (Boyd et al., 1989), shows that VCRs more often spread quickly in nations with highly controlled television systems because individuals tend to use VCRs to add what they find missing from television: entertainment, critical news, programs in their own language, sexual content, or whatever else is missed. Quantitative research by Straubhaar and Lin (1988) conversely showed a negative correlation between diversity in television systems (in terms of number of channels available) and VCR penetration rates, when nations are compared.

Social class modifies these general trends in two ways. First, as noted previously, wealthier people have more access to VCRs, particularly in poorer nations, even though VCRs are becoming general public media in the more industrialized countries and the oil-rich countries. Second, different social classes tend to have distinct patterns of viewing interests, which can lead members of a given class to use VCRs to seek material that may not be broadcast on television. Putting these two trends together, elites and upper middle classes in many—if not most countries—may now use VCRs to see material of interest to them that may not be of sufficient interest to appear on broadcast television, even in relatively diverse TV systems.

A specific example from the author's own research in several Latin American countries is the trend among elites there to use VCRs to obtain imported feature films, particularly from the United States. Although feature films from the United States are popular on broadcast television, they must compete with other kinds of programming, such as news, *telenovelas* (prime-time serials), variety shows, comedies, national or regional music shows that are very popular with middle and lower classes.[1] Because elites are offered less U. S. film programming than they would prefer on television, they seem to buy and use VCRs to get it. On the other hand, middle and lower classes, being relatively content with what is on television, see a low marginal utility for VCRs.

---

[1] Based on categorical analyses of television listings by Straubhaar (1984) and Chen (1987).

What creates this situation are a series of developments in Latin American (and other) television industries. National television production has indeed remained at very low levels in a number of countries, ranging from the English-speaking Caribbean to smaller nations of Europe, Asia, and Africa (Varis, 1984). Nevertheless, a number of countries, particularly in Latin America, but including a number of the larger and more prosperous countries of Asia and Africa, have rapidly increased the productive capacity of their television systems. The Dominican Republic, an island of roughly 6 million people, has gone from producing roughly 25% of its programming in the early 1970s to around 60% in 1988.[2] The Arab Gulf states have increased their television production, as have several nations of East Asia, and others.

Changes in national television production really deserve a more extensive and separate treatment, but can be traced to several elements. The costs of television/video production, recording and editing equipment have dropped very rapidly with solid state and computerized electronics. Furthermore, production equipment is now more portable and flexible, allowing a number of new uses, and in many cases requiring less training to operate. More trained people are available, after years of television experience. Both advertising-based commercialization and, in some countries, the success of noncommercial systems, have enabled a number of countries to finance more production, as well. Finally, more low-cost genres or program formats have been developed, such as the serial/soap opera, variety shows, live music, simple comedy, and talk shows, which are popular as well as relatively cheap and easy to produce.

These production developments have done several things that are relevant to social class and to video use. The first, notable for a number of years in much of Latin America and some other countries, is to reduce the amount of U. S. programs on television. This tends to please the general audience, but as noted earlier, is not necessarily in the viewing and cultural interests of the elites.

However, increased production capacity also enables the segmentation of the audience. This trend is quite evident in the United States and is becoming evident in Latin America. In Brazil and Bolivia, for example, there are channels clearly aimed at lower middle and lower class audiences, and in Brazil and Mexico, at least, there are channels clearly aimed at the upper middle class and above. This actually tends to retard VCR use, as an increased number of groups find that television generally meets their needs, and consequently the marginal utility of VCRs remains low, except for those people who are sufficiently wealthy that marginal utility judgments do not have to be made.

There are a few other TV/video industry developments that also affect

[2] Based on the categorical analyses of television listings by Straubhaar (1984) and Chen (1987).

the context for VCR use. In some countries, television industries are consciously segmenting their programming to target specific audiences to avoid losing them to VCRs. Malaysian television has recently added a second channel with some Chinese-language programs to try to draw a disaffected Chinese community back to watching government television channels, for example. Saudi television has similarly added a second channel with more entertainment to draw audiences back from VCR dependence (Boyd et al., 1989).

As in the United States and Canada, segmentation in some countries is being rapidly increased by the advent of cable television. In some cases, cable is actually brought in and operated by television owners, such as Televísa in Mexico, to insure a share in profits from video proliferation. More often cable TV operates in competition with both television and VCRs. In the United States and other relatively wealthy countries, there is a marginal utility to both television and cable viewers in owning a VCR for time-shifting and library building (Greenberg & Heeter, 1988; Levy, 1987). Where people cannot afford a VCR for the convenience of time-shifting, cable and VCRs tend to compete as alternative means of adding diversity to television.

One response, particularly by commercial television industries, is to increase their overall vertical integration by adding video rentals for VCRs. As noted earlier for cable TV, some commercial broadcasters, including Televísa (the example of television diversifying into cable ownership in the Third World), are moving to control video distribution, as well as overall film distribution. The move into vertical integration of television, video and film distribution by Televísa in Mexico, Venevision in Venezuela, and TV Globo in Brazil is somewhat parallel to similar moves toward cross ownership in cable TV, film production/distribution, and television broadcasting in the United States (see Komiya & Litman, chapter 2, this volume).

In countries where film producers are unable to control pirate distribution of their films of illegal video copies, there is a strong push to gain legal protection. This is as necessary in India or Egypt as in the United States or Great Britain. Some Third World countries in Latin America and Asia are indeed cracking down on piracy to protect domestic film industries and to avoid offending international interests such as the Motion Picture Export Association of America (Boyd et al., 1989).

INDIVIDUAL CONTEXTS, SOCIAL CLASS USES, AND EFFECTS

We have been looking at social class and VCR use in series of contexts. At the next level down, one could speak of smaller, almost individual contexts that are closely related to socioeconomic class: income levels, neighborhood, local systems of access to VCRs, occupations, education, family networks, ethnicity, language, and various subcultures. All of these affect both access to VCRs and the interests and values that determine use of VCRs.

Sometimes those who look at social class as a principal cause are divided between those who look at objective criteria, such as income or occupation, whereas subjectivists tend to look at common interests and consciousness of common interests (Mombert, 1930). I consider both important and look at both here, tending to link objective class indicators to questions of access and subjective conditions of class to interests.

Income, neighborhood, and family networks all help determine individuals' access to VCRs. Other aspects of class, such as occupation, education, and consumption orientation and ability help determine interests in whether to acquire and how to use a VCR. Some related demographic issues such as age, ethnicity, and language, and membership in key subcultures also affect VCR use. There seem to be several differences related to class in how and why VCRs are acquired and used in industrialized, oil-rich, newly industrializing and poor societies, that are explored further later.

For theorists based in dependency or other neo-Marxist analyses, social class stratification is critical in understanding cultural processes such as television production, television viewing, and the consequent roles of VCRs. This is not only true for the Third World. Bourdieu (1984) used social class among both producers and consumers as a principal analytical tool for understanding modern culture in France and other Western analysts have employed similar frameworks.

Defining social class is a complex task. In most definitions, *social class* is based in large part on disparities in income distribution, with variations and gradations between rich, middle class, working class, and poor. Income distribution is also a convenient variable because it can be measured and quantified. In setting the context for examining class differentiated use of VCRs, particularly at the international level, this chapter has thus far concentrated on income distribution questions. This is because of the clear relationship between income and VCR acquisition, on both an individual and a macro/societal level in all but the wealthiest countries. Income itself seems to be primary in determining who can afford a VCR, thus in opening or limiting access to them.

Within most societies, however, social class has a more complex and variable set of definitions, which will also be relevant to shaping both access to and interest in a new medium like VCRs. For many, particularly those building on Marxist theories, a person's occupation and role in the relations of production are perhaps more critical than income per se in determining their social class. Marx created a dichotomy between those who owned the means of production, whom he saw as becoming increasingly wealthy, and the workers and peasants, whom he saw as becoming impoverished (1953). A subsequent problem is accounting for groups that do not fit this original scheme. The principal problems for occupation-based definitions of social class in industrialized countries are controversies over the roles of professional classes who are wealthy and even powerful but do not own means of production, a series of layers of middle classes, working classes that range from well-off to impoverished, and an underclass of those who have no regular employment, but are usually sustained by a welfare system (Dahrendorf, 1959).

In many Third World countries, the same groupings and relationships trouble Marxist definitions, but the relative emphasis shifts to how to deal with fragile but real middle classes, working classes, agricultural workers, subsistence farmers, the very large body of those who are not regularly employed in the formal economy but do in fact engage in economic activity in the countryside or the urban informal economy, and those who are truly marginal and have no income at all (because few Third World states have effective welfare systems). As an example of adapting Marxist analysis to Third World conditions, in the Dominican Republic, Bosch (1986) called most of those in the informal sector "poor petit bourgeoisie" because they are effectively self-employed and tend to own some small productive equipment. However, calling an economically marginal peddler with a pushcart a poor petit bourgeois seems to stretch the definition of bourgeois pretty far.

A person's occupation does seem to relate to cultural choices, as Bourdieu (1984) indicated for music in France. He provided several examples where taste in music is clearly separated by respondents' occupational background. Similarly, Straubhaar (1989) found that occupation separated those who preferred imported versus local television in the Dominican Republic. Those in upper class occupations, such as managers, store owners, and professionals, tended to prefer imported programming from the United States over local or Mexican entertainment programs, in contrast to blue-collar workers, household servants, and peddlars, who preferred the latter.

Bourdieu also emphasized the importance of education in defining social class. He noted that education tends to build on and reinforce family background. In analyzing social distinctions in cultural preferences or tastes, he discussed "cultural capital," a set of knowledge and esthetic orientations that tend to come from family background, formal education, and information socialization associated with schooling, particularly at the university level.

Aspects of social class interact and modify each other. Within classes (as indicated by occupation), Bourdieu (1984) observed differences in cultural choices (in this case, about which photographic subjects are most pleasing) that are indicated by level and type of education, (pp. 35–39), even though he tended to see family background as more determinative of taste than education. Within an educational group, those from more privileged class family backgrounds will know more about "higher" culture, particularly those aspects, such as film directors, avant-garde music, and so forth, which are not taught in school curricula (Bourdieu, 1984, pp. 63–65.)

Given that VCRs greatly increase individuals' choices about what to watch on "television," a variety of groupings of interest can become significant. For instance, differences in ethnicity and language can produce minority cultures or cultural enclaves that see the VCR as one means of selecting nonbroadcast programs that speak directly to them and their interests.

In some cases, language is the critical issue. If there are larger groups elsewhere producing film or television programming in a language, minority language communities outside the home culture can now use VCRs to watch it. Virtually every sizeable U. S. city, as well as many smaller college towns, now have Chinese, Korean, Indian, Arabic, or other ethnic grocery stores or restaurants that also rent videotapes from "home." Typically, both films and a variety of television programs can be found (Dobrow, 1986). Hispanics in many U. S. communities are served by broadcast or cable channels in Spanish but many Hispanic VCR rentals can also be found. Some have raised the question as to whether the easy availability of foreign language videotapes will reduce the speed with which immigrants to the United States will learn English and acculturate (Yoo, 1987).

Ethnic use of VCRs in some other countries is even more striking. For example, in Malaysia, the ethnic Chinese are relatively wealthy and control a good deal of the economy despite being a small group. Partially to compensate, government-controlled television broadcasting has emphasized development-oriented programming in the Bahasa Malay language, as well as the Islamic religion (Boyd et al., 1989). In response, the Chinese essentially have used VCRs to avoid Malaysian broadcast television and watch Chinese videotapes from Taiwan, Hong Kong or elsewhere (Lent, 1985; Tan, 1987). It is worth noting that the Malaysian Chinese community's use of VCRs represents both a language/ethnic division and a social class division, in that the Chinese largely have the class position and money to afford access to VCRs.

Other subcultures may see VCRs as a means of tailoring television viewing to their own tastes. Rock music fans in many countries now build libraries of music videos, particularly if they are not prominent in either television or cable programming. Brazil has late night music programs on broadcast television that encourage fans to tape them for the videos.

Political groups, religious groups, unions, and other organizations are beginning to see VCRs as a means of both creating and distributing alternative political, economic, religious, or cultural messages. This is particularly notable in Latin America, so far, but has also been reported in India, other parts of Asia, and Africa (Boyd et al., 1989). Many of these alternative VCR projects take on a specifically class-conscious character, as with union, Catholic Church, and neighborhood video projects in Brazil designed to raise problems and to organize for alternative solutions (Festa & Santoro, 1987).

VCRs AND CLASS IN INDUSTRIALIZED SOCIETIES

In industrialized societies, classes are, at least by comparison with the Third World, less radically stratified. In fact, a major challenge for Marxist theoreticians has been to explain how middle, working, and agricultural classes have been coopted into accepting the hegemony of the industrial capitalist system (Gramsci, 1971). Although major income and other disparities do exist, in a number of countries majorities of individuals or households can afford something like a VCR, for which the world base price is at this writing about $200. So class is less of a determinant of access. Even so, neighborhoods and their relative density of rental shops/clubs affect access, as do collections of videotapes by friends, relatives, and neighbors. In fact, beyond access, group viewing of VCRs may well have strong social utility, particularly among some ethnic groups or classes.

In less-stratified societies, within-class differences seem to be more determinative of why and how VCRs are used. Considerable weight shifts to demographic variables like gender, age, and ethnicity, to interest groups (martial arts or aerobics students who watch and work out to appropriate videos), subcultures (like various rock music-oriented subcultures who watch their preferred music videos) or smaller interpretive communities. There are also weights on video use from family environments or household patterns, and not least, uses and gratifications at the individual level (Levy & Windhal, 1984).

In many of the industrialized societies, as well as oil-rich societies, such as Saudi Arabia, and newly industrializing societies, such as South Korea and Taiwan, VCRs are becoming a majority complement to broadcast television and where available, cable or satellite TV. In the oil-rich nations or newly industrializing nations, where television is less diversely developed, VCRs seem to play a greater role in the substitution of alternative content, frequently imported from other countries or from local or regional film industries, for what is broadcast. In both industrialized, oil-rich and newly industrializing nations, VCRs are also serving as tools for meeting minority groups' language and cultural needs. Even in the United States, Dobrow (1986) has observed that ethnic language groups frequently use VCRs to play material from "home."

SOCIAL CLASS AND VCR USE IN HIGHLY STRATIFIED SOCIETIES

In highly stratified societies, the differences between classes are more significant, including VCR access and use. This includes most of the Third World, even many nations that are considered to be newly industrializing, like Brazil. Although almost all Third World nations now have nascent middle and working classes, even people in these classes are hard pressed to afford a VCR, particularly because many Third World governments boost VCR prices by imposing importation limits, tariffs, and taxes on them (Boyd et al., 1989). Still, despite income limits, in many Third World countries, increasing numbers of those in urban areas or even villages have periodic access to VCRs via neighborhood bars, video clubs, or via friends, relatives, and neighbors. In India, for example, increasing numbers of people are setting up video clubs to show pirated versions of popular Indian movies and videos are shown to attract people to bars, public events and even competing lines of intercity buses, which show videos enroute. Hence, many people see video at least on occasion, including many who seldom—if ever—see broadcast television (Boyd et al., 1989).

The issue of access to VCRs by group viewing, commercial video "theaters," and the growth of videocassette recorder rentals complicates an already difficult problem of finding out who has VCRs. Penetration is usually calculated by known VCRs as a proportion of TV households, but in the Third World, sales figures are suspect when smuggling is widespread and relatively few surveys are conducted, except in Latin America and parts of Asia where consumer surveys are becoming both more common and more reliable.

In most Third World countries, VCRs are seen as an alternative to broadcast television, rather than a complement. However, in some countries of Africa and the Caribbean that have not yet begun broadcast television, VCRs are the means of entry for "television." The same is true of remote rural areas, such as the highlands of Peru or Bolivia where urban-originated television broadcasts may not reach and where television arrives either via VCR or satellite dish (Boyd et al., 1989). In these situations, social class' economic aspects tend to serve as a determinant of who gets access to television via VCRs or satellite dishes.

In the Third World as well as the First World, social class is an indicator of major types and levels of needs, as well as access, per se, to media such as VCRs. Despite VCRs' cost, the other aspect of marginal utility is the perceived need for entertainment and/or information beyond what the local or national television system provides. Many broadcasts simply lack either information or entertainment to such a dramatic degree that expending a remarkably high proportion of scarce resources to acquire the means to substitute something else for what is broadcast begins to seem worth the high marginal cost.

In some relatively wealthier communities or classes, part of the appeal, or marginal utility, is the status need to own a VCR. Boyd (1989) noted that in Egypt, middle-class families are embarrassed to actually go to movie houses because it indicates that they do not have a VCR with which to watch the film (usually in pirated version) at home.

SOCIAL CLASS, INTERESTS, AND VCRs IN LATIN AMERICA

One of the emerging conclusions of a good deal of current research on media and culture in Latin America is the strong role of class divisions in determining cultural uses and practices. Most researchers in Latin America, particularly those operating from the dependency theory perspective, would probably agree that elites and upper middle classes are increasingly internationalized. Marxist and Neo-Marxist theories have long predicted an internationalization of the bourgeoise. Further, it was expected that these elites would tend to dominate the process of hegemony or cultural consensus building and impose their own internationalized values on the lower classes as part of drawing them into the international capitalist system (Salinas & Paldan, 1976).

However, Barbero (1988) and Canclini (1988) noted that working class, artisan, and peasant cultures are not behaving as expected under industrialization. They are not necessarily following the elites or upper middle classes, but rather maintaining their own cultures, which frequently operate in opposition or resistance to the elites. Struggling middle and lower middle classes are in perhaps the most ambiguous position. They seem to be torn between emulating elites and maintaining ties with popular culture.

These class differences relate to television viewing, which then sets the stage for VCR use. If elites and upper middle classes are internationalized in work, travel, education, learning of English, and so on, then one might well expect them to come to prefer U. S. or other imported television and films. Ongoing empirical research, both a review I have done of ratings data in Brazil, Dominican Republic, and Venezuela and sociological and ethnographic interview studies in Brazil and the Dominican Republic (Kottak, 1989),[3] tends to show that television viewing is in fact divided by social class. Elites and upper middle classes do tend to prefer U. S. films and series, even though they do like national news and interview programs. Lower classes tend to prefer national comedies, music, *telenovelas*, and variety shows, or similar programs from other Latin American countries. In fact, music, *telenovelas*, and news tend to be popular with the broad general public (Rogers & Antola, 1985), but detailed analysis shows that elites are frequently disenchanted with them and prefer imported television entertainment (Straubhaar, 1989).

Within this general framework of class preferences for television in Latin America, there are some significant differences between countries in both economic and television systems contexts and these differences have profoundly affected VCR acquisition and use. In both Brazil and the Dominican Republic, VCR acquisition and use has been limited by economic restraints and television systems that tend to retain viewer interest and loyalty.

Brazil had relatively low VCR penetration (under 4% of television households) until 1986, when it rose to about 6%. VCR prices have been kept high by import limits and tariffs and, overall, incomes have been poorly distributed and limited despite a growth in both working and middle classes. Perceived marginal utility of VCRs also seems to have been low in light of the television system's performance, particularly its production of national entertainment programming and its segmentation of the audience. There is a major network (TV Globo) that produces 10–12 hours of programming daily and dominates the general audience, with a 60%–80% share of households watching TV.[4] Two networks target the upper and upper middle class, because ratings show that that group is less satisfied with TV Globo, and another network targets the lower

[3] I have examined questions about use of imported versus national television programs in two studies in the Dominican Republic, in Santo Domingo: a random sample survey of 255 adults in 1986 and an ethnographic interview study with 125 interviews in 1987, based on a snowball sample in poor, working-class, lower middle-class and upper middle-class neighborhoods. Kottak and colleagues have conducted 1,200 interviews in villages in the Brazilian states of Bahia, Para, Parana, São Paulo, and cities in Rio de Janeiro and São Paulo.

[4] Audience information cited here comes from the major Brazilian ratings firm, IBOPE. I have examined ratings data for 1 week every 2 years, 1963–1987.

middle and lower class, receiving high ratings in those groups. So many of the audiences' interests are covered by the broadcast television system, leaving a low marginal utility for VCRs. VCRs increased during an economic boom in 1986, rising to reach 6% of television households, but sales have stagnated in subsequent economic crises. VCRs are most popular with the upper and upper middle classes and are primarily used for watching U. S. movies,[5] which seems to go along with a class division of television viewer interests and VCR use.

The Dominican Republic has even lower average incomes and worse distribution of income, so for economic reasons, access to VCRs has been and remains limited. Santo Domingo had five broadcast television channels in 1985 and seven in 1988, so most viewers have some options. Furthermore, unlike Brazil, Santo Domingo and other major Dominican cities also have cable television systems that pirate 12–14 U. S. satellite cable channels and distribute them in upper and upper middle-class neighborhoods, so even among elites, VCRs have to compete with cable, which costs around $8 per month for HBO, Cinemax, Disney, Movie Channel, and 8–10 other channels. VCR penetration was 2% of television households in 1984 and remained under 3% in 1988. Interviews that I conducted in 1984 and 1988 show that a number of VCR owners obtained them for status reasons and did not use them much, since a rental market has never really developed, beyond a handful of shops.

By contrast, VCRs diffused much faster in Colombia and Venezuela, where incomes tended to be somewhat higher than in Dominican Republic and even Brazil, particularly in the early 1980s. Neither the Colombian nor the Venezuelan government made any particular effort to limit or tax VCRs, which kept prices relatively low. In addition, VCRs acquired a more widespread and visible consumption object status in both Colombia and Venezuela, particularly for the middle class (Bibliowicz, 1982). VCR acquisition seems to have slowed in Venezuela, along with the economy in general, but 31% of Venezuelan television households had VCRs by 1984, compared with 18% in Colombia (Boyd et al., 1989).

The television systems in both Colombia and Venezuela also had fewer channels than Brazil or Dominican Republic, probably indicating lower diversity, which seems to have raised the marginal utility of VCRs in Colombia and Venezuela. Social class division in VCR acquisition and use seems to have been less extreme, but VCRs still seem to have remained limited to the middle class and up, according to consumer surveys in 1984-1985. Use, as indicated by interviews with rental shop owners in Venezuela, seems to have been predominantly for renting feature films

[5] VCR use conclusions are based on weekly surveys by *Folha de São Paulo* of video rental shops to determine the most popular rental videos.

from the United States, which fits with an elite-upper middle-class pattern of use, as well as a pattern of middle-class imitation of elite and upper middle-class consumption.

Overall, VCR use in these Latin American countries seems stratified by class. Particularly while costs remain relatively high, compared to incomes during the overall regional stagnation of the mid- to late 1980s, lower and lower middle-class audiences seem to resolve the question of utility in favor of continuing to watch broadcast television. Equally striking, however, elites do seem to be acquiring VCRs and using them to watch U. S. feature films.

CONCLUSION

Most observers would agree that VCRs are personalizing television, permitting viewers to tailor their video consumption more narrowly to their own particular interests. What is only partially visible in the United States is that in more stratified societies, VCR access and use can reveal much deeper and more systematic cleavages in culture between groups. In particular, this chapter has argued that those cleavages are frequently along the lines of social class, as well as language, ethnicity, and subcultures.

In more stratified societies, social class interacts with both economic contexts (at the world and national levels) and with the particular context of the national television system to affect access to and use of VCRs. The economic context essentially tends to determine who can afford access to VCRs, but that is somewhat dynamic. Although both the Arab oil states and the rapidly industrializing states of East Asia are counted among the Third World, several have high or rapidly increasing incomes and rapid VCR diffusion. Similar economic growth could expand VCR diffusion elsewhere, although such diffusion can start and stop with economic cycles, as in Venezuela.

The world capitalist economy also directly affects the roles and interests of the different classes, particularly in economically dependent societies. Most specifically their role in the world economy tends to internationalize elites and upper middle classes' cultural interests, which they tend to use VCRs to follow up on.

The national television system context affects VCR acquisition and use by meeting the interests of some groups and classes and not meeting the needs or interests of others. In particular, the success of some national television systems in creating extensive national programming meets the interest of most of the viewing audience, who at least in the countries emphasized here definitely seem to prefer national programming overall. However, that "nationalization" means that those classes with more internationalized tastes seem likely to turn to the VCR (and the cable and satellite dish) to obtain more imported programming, usually from the United States. The other two main instances in which national television programming orientations drive VCR use are when television is oriented to development programming and urban middle classes and elites crave more entertainment, and when broadcasters' language policies, implicit or explicit, excludes some groups getting the ethnic/language group programming they wish. In all three cases, the separation between class and/or ethnic group cultural interests and values is being facilitated by the VCR, which is somewhat ominous for the solidarity of those societies. In the process of giving individuals and groups more choices, the VCR seems likely to reduce the cultural bonds and common references and information that tie societies together. There may be a cycle of cultural preference in which elite preferences do trickle down or popular tastes be appropriated by elites, but the current tendencies do not seem to indicate that.

In summary, VCRs can and will be used to express class needs as well as personal interests. In fact, VCR choices probably are a good index of class and other identities. Some classes' needs will be met by broadcast television; it is likely that elites and minorities will be least satisfied. Those groups will likely use VCRs and other new media to satisfy their interests, breaking down the cultural common ground that broadcast television has been hoped to encourage. The abilities of different classes to use VCRs to meet their needs will be powerfully affected by income level and distribution, an aspect of different patterns of stratification across nations.

REFERENCES

Barbero, J. M. (1988). Communication from culture: The crisis of the national and the emergence of the popular. *Media, Culture and Society, 10*, 447–466.

Bibliowicz, A. (1982, June 24). La necessidad de las opciones multiples [The necessity for multiple options]. *Tele Revista*. (Bogotá).

Bosch, J. (1986). *La pequeña burgesia en la historia de la Republica Dominicana* [The petit bourgeoise in the history of the Dominican Republic]. Santa Domingo: Editora Alfa y Omega.

Bourdieu, P. (1984). *Distinction: A social critique of the judgement of taste*. Cambridge, MA: Harvard University Press.

Boyd, D. A. (1988, May). *The videocassette recorder and the dissemination of western cultural and political information in the USSR and soviet countries*. Paper presented at the International Communication Association Conference, New Orleans, LA.

Boyd, D. A. (1988, May). *The videocassette recorder and the deterioration of the Egyptian film industry.* Paper presented at the International Communication Association Conference, San Francisco, CA.

Boyd, D. A., Straubhaar, J. D., & Lent, J. A. (1989). *Videocassette recorders in the third world.* New York: Longman.

Canclini, N. G. (1988). Culture and power: The state of research. *Media, Culture and Society, 10*, 467–498.

Cardoso, F. H., & Faletto, E. (1979). *Dependency and development in Latin America.* Berkeley, CA: University of California Press.

Chen, S. (1987). *The flow of television programs from the United States to Latin America: A comparative study of two week's television schedules in 1973 and 1983.* Unpublished master's thesis, Michigan State University, East Lansing, MI.

Chilcote, R. H. (1984). *Theories of development and underdevelopment.* Boulder, CO: Westview Press.

Dahrendorf, R. (1959). *Class and class conflict in industrial society.* Palo Alto, CA: Stanford University Press.

Dobrow, J. R. (1986). *The social and cultural implications of the VCR: How VCR use concentrates and diversifies viewing.* Unpublished doctoral dissertation, University of Pennsylvania, Philadelphia, PA.

dos Santos, T. (1978). *Imperialismo y dependencia* [Imperialism and dependency]. Mexico, DF: Ediciones Era.

Evans, P. (1979). *Dependent development: The alliance of multinational, state and local capital in Brazil.* Princeton, NJ: Princeton University Press.

Festa, R., & Santoro, L. (1987). Policies from below: Alternative video in Brazil. *Media Development, 34*(1), 27–30.

Ganley, O., & Ganley, G. (1988). *The political implications of the global spread of videocassette recorders.* Norwood, NJ: Ablex.

Gramsci, A. (1971). *Selections from the prison notebooks of Antonio Gramsci* (Q. Hoare & G. N. Smith, Eds. and Trans.). London: Lawrence & Wishart.

Greenberg, B., & Heeter, C. (1988). *Cable viewing.* Norwood, NJ: Ablex.

Kottak, C. (1989, April). *Presentation of results of an ethnographic study on the effects of television on four villages and two towns in Brazil.* Paper presented at the Third Conference on Latin American Popular Culture, Michigan State University, East Lansing, MI.

Lent, J. (1985). Video in Asia: Frivolity, frustration, futility. *Media Development, 32*(1), 8–10.

Lerner, D. (1958). *The passing of traditional society.* New York: The Free Press.

Levy, M. R. (1987). The VCR age. *American Behavioral Scientist, 30*(5).

Levy, M., & Windahl, S. (1984). Audience activity and gratifications: conceptual clarification and exploration. *Communication Research, 11*, 51–78.

Marx, K. (1953). *Das kapital* [The capital]. Berlin.

Mattelart, A., & Schmucler, H. (1985). *The new communication technologies: Freedom of choice for Latin America.* Norwood, NJ: Ablex.

Mody, B. (1989, May). *The commercialization of TV in India: A research agenda for cross-country comparisons*. Paper presented at the International Communication Association Conference, San Francisco, CA.

Mombert, P. (1930). Class. In E. R. A. Seligman & A. Johnson (Eds.), *Encyclopedia of the social sciences*. New York: Macmillan.

Rogers, E. (1976). Communication and development: Critical perspectives. *Communication Research, 3*(2).

Rogers, E. M., & Anatola, L. (1985). Telenovelas. *Journal of Communication, 35*(4), 24–35.

Salinas, R., & Paldan, L. (1976). Culture in dependent development. In K. Nordrenstreng & H. Schiller (Eds.), *National sovereignty and international communication*. Norwood, NJ: Ablex.

Sarti, I. (1985). Communication and cultural dependency: A misconception. In E. McAnany, J. Schnittman, & N. Janus (Eds.), *Communication and social structure*. New York: Praeger.

Sorokin, P. (1928). *Contemporary sociological theories*. New York: Harper & Row.

Souki de Oliveiria, O. (1989). Media and dependency: A view from Latin America. *Media Development, 36*(1), 10–14.

Straubhaar, J. D. (1984). Brazilian television: The decline of U. S. influence. *Communication Research, 11*(2), 221–241.

Straubhaar, J. D. (1986, August). *Cultural dependency in Latin America*. Paper presented at the International Communication Association Conference, New Delhi.

Straubhaar, J. D. (1989, May). *The impact of cable TV in the Dominican Republic*. Paper presented at the International Communication Association Conference, San Francisco, CA.

Straubhaar, J. D., & Lin, C. (1988). A quantitative analysis of the reasons for VCR penetration worldwide. In J. Bryant & J. Salvaggio (Eds.), *Media use in the information age: Emerging patterns of adoption and consumer use*. Hillsdale, NJ: Lawrence Erlbaum Associates.

Tan, S. Y. (1987). *Broadcasting system in Malaysia: The influence of videocassette recorders*. Unpublished master's thesis, Michigan State University, East Lansing, MI.

Varis, T. (1984). The international flow of television programs. *Journal of Communication, 34*(1), 143–152.

Wallerstein, I. (1977). *The capitalist world economy*. London: Cambridge University Press.

World Bank. (1989). *World development report*. Washington, DC: Author.

Yoo, E. (1987). *VCR use by Korean minorities in the United States and its cultural implications: Am empirical study*. Unpublished master's thesis, Michigan State University, East Lansing, MI.

III

The Relationship of VCRs to Individual Behavior and Use Patterns: Individual Expression, Collective Identity, and Social Patterns

8

VCR Libraries: Opportunities for Parental Control

Katharine E. Heintz
University of Illinois—Urbana–Champaign

Since the introduction of the VCR into the United States in the 1970s, media critics and social analysts have disagreed over its potential impact on Americans' leisure time. Predicting a saturation level of 10%–14% of U. S. homes, Agostino, Terry, and Johnson (1980) concluded that VCR use was "unimaginative" and conditioned by audiences' broadcast viewing habits. They predicted that VCR penetration would be limited to up-scale households and used to supplement their already existing media technologies. Similarly, *Broadcasting* magazine in 1981 forecast a minimum penetration of 25% by 1990. In an October 1988 issue of *Time* magazine, however, Richard Zoglin indicated that VCRs are now present in over 60% of American households, making it the most quickly disseminated new communications technology in history (Levy, 1987).

This dramatic increase in the number of American households with VCRs during the 1980s has caused mass media researchers and the broadcast and cable industries to take notice. Although the networks claimed 92% of the prime-time viewing audience in 1978, today's figures suggest the network share of the audience is only about 70% (Zoglin, 1988). Certainly this drop cannot be attributed solely to the expansion of VCR technology. The number of independent television stations and cable stations increased dramatically during the 1980s as well, providing more alternatives to network programing.

As their audience share decreased, networks became interested in what the viewers were watching. Network executives breathed a sigh of relief when it was consistently discovered that the most common use of the VCR was for time-shifting regular broadcast content to a more convenient viewing time (Agostino et al., 1980; Levy, 1980a, 1980b, 1983). Rather than losing audience members to prerecorded videotape content, broadcasters were garnering an even greater audience for some programs. For example, soap opera viewing among school-aged females increased because programs could be taped and watched after school (Greenberg & Heeter, 1987). Similarly, programs airing simultaneously on different channels could both be viewed through the process of time-shifting (see Agostino et al., 1980; Levy, 1980a, 1980b), thus increasing the audience size for broadcast programs.

Not all VCR research has been good news for broadcasters, however. An increasingly popular reported use of VCRs is for playing prerecorded videocassettes either rented or purchased from a video retailer (Levy, 1980a; Rubin & Bantz, 1987; Wartella, Heintz, Aidman, & Mazzarella, 1990). According to Barcus and Hughes (1988), children's videos make up approximately 20%–30% of the total video market, second only to feature films. The authors also reported that a higher proportion of children's videos are sold than general audience films or videos. Indeed, this dramatic growth in the number of videocassettes available for rental or purchase was one of the major reasons cited by the Federal Communications Commission (FCC) in its 1983 decision not to require broadcasters to air age-specific programming for children. Claiming "there is no national failure of access to children's programming that requires an across-the-board, national quota for each and every licensee to meet" (FCC, 1983, p. 648), the FCC considered VCRs as full participants in the household media environment.

The FCC drew their conclusions when approximately 10% of U. S. households owned VCRs (Ganley & Ganley, 1987). In the intervening years, VCR ownership in this country has increased sixfold. And the videocassette market has grown into a $7 billion industry (Mayer, 1988), with over 45,000 rental and sales outlets in the United States (Mayer & Sweeting, 1987). According to the 1989 *Channels Field Guide*, total factory sales of videocassettes was over 3.7 billion tapes in 1988 (Mayer, 1988). With the decrease in the cost of videotapes, more viewers can afford to purchase their own and begin compiling home videocassette libraries. Indeed, industry analysts predict that revenue gained from sales of prerecorded videocassettes will surpass that gained from video rental by 1990 (Mayer & Sweeting, 1986).

Because of the expanded popularity of this technology, then, mass media researchers have to ask how this new medium affects our other leisure-time pursuits. And recognizing that households with children are more likely to have VCRs than childless households (Greenberg & Heeter, 1987), we must ask what effects, if any, this technology has on children. More specifically, is the VCR used to provide children with a wider range of content than is available on broadcast or cable television, or is it used to limit the range of content available to child viewers through repeating or replaying standard television fare?

There are two basic issues under consideration in this examination. On the one hand, if VCRs are used to increase the variety and diversity of content viewed on television, then children growing up in VCR households could be better prepared for school or other learning situations than children in non-VCR households. The knowledge gap could be widened by the purposive use of VCRs. On the other hand, if VCRs are used to limit the variety and diversity of content viewed on television, then we could find two very different results. First of all, limiting VCR use to showing programs that were either originally broadcast on TV or cable, or resemble standard commercial television fare (i.e., animated toy-related programs) could contribute to the knowledge gap through decreasing child viewers' exposure to a variety of higher quality media fare. Conversely, children's VCR use could be limited to watching only specific program genres that cannot be easily found in the normal commercial broadcast or cable schedules (i. e., instructional or educational programs, science programs, culturally specific programs). In this situation, the VCR could fulfill a narrowcasting function (predicted by Levy, 1980a), allowing individual viewers and families to tailor their viewing selections to their special interests and values.

Both of these issues have been addressed in earlier research on videotapes available for rental. Dobrow (1989) provided evidence of immigrants in the United States using videotapes to view material in their native languages instead of, or in addition to, over-the-air TV. Her results highlight the important narrowcasting potential of the VCR. This practice can work to enhance cultural plurality through the provision of opportunity to select content that presents an alternative to "mainstream" American media fare. Conversely, Wartella et al. (1990) examined the diversity of videotape genres available for rent (or library loan) in a small Midwestern community. An analysis of video retailers' collections indicated that most of the children's content provided by video outlets was virtually identical to that offered by the local commercial broadcast and cable outlets. So, for the children in this community, videotapes provided more of what was available to them on TV. Unlike Dobrow's evidence of a narrowcasting function of VCRs among her sample, Wartella et al. showed that for their sample, VCRs are used to further increase the audience for broadcast or broadcast-type fare.

The importance of variety of videocassette selection for children's socialization is clear. If videocassettes provide a picture of the world that is consistent with that continually pictured on commercial television, then tapes will serve to reinforce the "mainstream" or hegemonic ideology. If, on the other hand, videocassettes provide alternative ideologies, child viewers may develop less stereotypic and more pluralistic ideas about the world. In other words, videocassettes can function as "emancipatory popular culture" (Kellner, 1987). According to Kellner, "Emancipatory popular culture challenges the institutions and way of life of advanced capitalist society . . . [It] subverts ideological codes and stereotypes, and shows the inadequacy of rigid conceptions that prevent insight into the complexities and changes of social life" (p. 489). In addition to the presentation of foreign cultures and alternative ideologies, videocassettes can be used by parents to encourage children's academic or social learning and creativity, or to provide nonviolent alternatives to broadcast cartoons. Thus, parents who are dissatisfied with the content of children's broadcast and cable programming may deliberately choose to assemble a library of alternative content for their children, if such content is available.

In this chapter, I propose to report a further analysis of the Wartella et al. data, looking specifically at home videotape collections, or libraries. Without denying that the variety of children's videotape genres available through retailers in the community mimics the selection provided by the local commercial broadcast and cable outlets, I believe that it is possible for interested parents to select from among those genres and develop a viewing diet consisting of both TV and video content that supports their values and attitudes toward television viewing. Therefore, I examine the relationship between videotape ownership and parents' uses of and satisfaction with broadcast television, cable television, and videocassettes. I also assess the link between parental control over children's television viewing and their provision of alternative viewing material via tape libraries. Finally, I consider the content of home video libraries to see if they perform a narrowcasting function, or if they serve to increase the audience for mainstream broadcast fare.

The term *library* refers to a collection of videocassettes purchased by the parents for their family. I use the term *library* rather than *collection* or *compilation* for two reasons. First of all, most videocassettes available for children are narratives. Whether one 60-minute long story or several short stories appear on a tape, a vast majority of children's videos are in the narrative form and can be described as talking and moving picture books. (Indeed, many are adaptations from picture books.) Second, the increased availability of children's videocassettes in book stores and libraries indicates that this form is considered akin to printed stories. Jordan (chapter 9, this volume) observed that families in her sample displayed their videocassettes on "bookshelves," along with their books, indicating that tape owners make the connection between print and video sources for narratives. Hence, in this chapter, *videocassette library* refers to a family's purchased collection of videocassettes.

The data for this chapter were collected in a case study of Champaign–Urbana, Illinois. According to the *Broadcasting/Cablecasting Yearbook* for 1987, Champaign–Urbana constitutes the 75th media market in the country. It is a moderately sized academic community, with slightly fewer than 100,000 inhabitants. The population is predominantly White, with about 10,000 Asians and Blacks. It is a fairly affluent community, with median family income in both of the twin cities above the national average. As would be expected in a college town, residents of Champaign–Urbana are also well educated.

The rationale for conducting a case study of one community lies in the nature of the phenomenon of interest. Although much commercial television fare is nationally distributed, cable and videocassette distribution is more local in character. VCRs are increasingly becoming more common in TV households, but the availability of videocassettes for rent or sale in any given community could limit the potential uses for VCRs. For this reason, it is important to examine smaller markets in addition to the major urban centers to assess whether all potential audience members are being served equally by the new communications technologies.

The data reported were collected through random phone surveys in the community in the Fall of 1987 and the Winter of 1988. Our final sample consisted of 214 completed interviews with parents who had at least one child under 12 living in their home.[1] To examine this issue of videotape libraries, the subsample of respondents who indicated that they owned at least one videocassette ($N = 144$) was divided at the median into "library owners" (10 or more videocassettes owned, $M = 39, N = 76$) and "nonlibrary owners" (less than 10 videocassettes owned, $M = 4$, $N = 68$). These two groups are compared on their responses to questions addressing their uses of and attitudes toward a variety of media, control of their children's media use, and content of their tape collections.

DESCRIPTION OF SAMPLE

Earlier research found that VCR households were more likely to be upscale and "media rich," that is, VCR owners are more likely to also subscribe to cable and pay cable services, own multiple television sets, and to own a home computer (Agostino et al., 1980; Greenberg & Heeter, 1987). The sample of Champaign–Urbana VCR owners, however, indicates that VCRs are present in homes from all levels of the socioeconomic ladder and with a variety of different media environments (see Wartella

[1] For a more complete description of the methods used, see Wartella et al. (1990).

et al., 1990). The diffusion of this technology has progressed past the initial stages so that we no longer see the pattern of early adoption with owners belonging to the more up-scale, informed segments of society. Now we are discovering VCRs in households without cable or computers, in both single-parent and two-parent families. The adoption rate of this technology indicates that it probably will play a part in the lives of a great majority of American children.

Demographically, tape library owners do not differ significantly from nonlibrary owners. I found no relationship between family's ethnicity or socioeconomic status (SES) and number of tapes owned by a particular household. Early adopters of the technology (those who purchased their VCRs before 1983) tend to have larger tape libraries than late adopters, but they own fewer children's tapes. This ownership pattern could reflect both the rapid growth of the video market in recent years as well as the fact that children's tapes are purchased more often than any other types of videos (Barcus & Hughes, 1988). Early adopters, then, have had more time to create their libraries, but a smaller overall variety of tapes from which to choose.

There was an overall difference in size of videocassette library by age and gender of child. Parents of older children and boys reported that they own significantly more tapes specifically for their children than parents of younger children and girls. Parents were also asked to identify the genres of tapes they have purchased for their children from a list of genres obtained from video merchants. When we look at the types of tapes purchased for older versus younger children we find that, although older children have more tapes to call their own, these are most often family or adult oriented. Younger children's tapes are more likely to be labeled as "children's" genres (i. e., animated, toy-related, or Disney). This viewership pattern is consistent with previous research that indicates that children in elementary school begin watching and often prefer family and adult-oriented programs (see Dorr, 1986; Wartella et al., 1990). An interesting finding in the data revealed that even though older children have more videotapes to call their own, younger children spend a significantly larger amount of time watching videos than older children ($t = 1.90, p < .06$).

If we look at viewing habits, we find that children from families with a library of videotapes watch less TV ($M = 14.6$ hours/week) than children from families with few or no tapes ($M = 18.2$ hours/week). These results could prompt a couple of different conclusions. Perhaps videotape library owners are also more restrictive of their children's TV viewing, limiting and number of hours their children watch. Or perhaps videotape library owners have a wider variety of media available in their homes to occupy more of their children's time.

To assess the variety of media available to children, the number of different types of media found in the home was summed to obtain an index of home media environments. Families with extensive video collections own significantly more types of media than families with more limited video libraries ($t = 2.55$, $p < .01$). Still, over 80% of both groups have at least one television hooked up to cable and approximately 20% of each group subscribe to a pay movie channel. Looking specifically at children's media available in the home, children from families with tape libraries have slightly more media they call their own than children in families without tape collections, and these families are twice as likely to subscribe to the cable Disney channel. The alternative activities hypothesis seems to be operative in this instance, with children dividing their time among different media when they are available, and spending more time with TV when they are not. It appears, then, that parents who are accustomed to providing media opportunities for their children are likely to incorporate children's tapes into the overall selection.

PARENTAL SATISFACTION WITH AUDIOVISUAL MEDIA

Parents report moderate levels of satisfaction with the quality of children's programming available on broadcast television, cable television, and in video stores. Library owners are slightly more likely to be satisfied with Saturday morning television programs than nonlibrary owners; library owners are also less likely to object to toy-related programs than nonlibrary owners. These results are surprising considering the highly educated background of the respondents. Educational level is significantly related to attitudes toward toy-related programming and MTV, however. College-educated respondents were significantly more likely to object to both toy-related programs and MTV than noncollege educated parents ($\chi^2 = 31.97, p < .001$). These attitudes did not appear to influence parents' purchase of videocassettes as alternatives for their children; the relation between parents' education level and size or content of children's tape libraries was nonsignificant.

All parents had high praise for both public television and cable offerings for children and many reported that they taped programs from these channels to keep in libraries for their children. From these responses, then, it appears that parents are not compiling videocassette libraries because they are dissatisfied with broadcast and cable children's programming. In fact, the majority of parents in this sample report purchasing blank videocassettes for the purpose of recording broadcast and cable programs and movies for their children.

PARENTAL CONTROL OF AUDIOVISUAL MEDIA

When we examine the variable of parental control of different media, there is no significant difference between the two groups on any of our measures. A control index was computed by summing all of the "yes" responses to questions about restriction of broadcast, cable, and VCR use. I found a small, yet significant, negative correlation between level of parental control and number of videocassettes the family owns ($r = -.163, p<.05$). It seems, then, that parents who restrict access to and selection of television and cable fare extend this limitation to alternative audiovisual content as well. It is those parents who are most likely to limit the amount and types of television and cable programs their children view who are also more likely to restrict the availability of alternative viewing matter through a small videocassette library.

Still, a majority of each group report that they are moderately to heavily restrictive of their children's viewing, especially younger children. Yet, in spite of their levels of television control, a majority of all parents report that they never or only occasionally watch broadcast or cable programs or videocassettes with their children. In her ethnographic analysis of family media behavior, Jordan (chapter 9, this volume) observed that parents feel less compelled to watch videocassettes with their children than broadcast or cable TV. The parents in Champaign–Urbana, however, do not seem to make this distinction, reporting control and co-viewing (or lack of) equally across this different media.

A statistical comparison of parent's control of television and VCR viewing shows significant consistency across the two media. Parents who limit the number of hours per day that their children may watch television also limit the number of video viewing hours per day ($\chi^2 = 15.17$, $p<.0001$); parents who limit television viewing to specific times during the day do so for video viewing as well ($\chi^2 = 19.15, p<.0001$). Similarly, parental co-viewing is significantly consistent (at $p<.0001$ for all comparisons) for television, cable television, and video programming.

Because we interviewed only parents, we have no check on the validity of their responses, and can only conclude that all the VCR owners in our sample exert some control over their children's media use. The above average education and income level of our respondents is consistent with higher levels of control (see Blosser & Heintz, 1987). However, these characteristics have also been shown to predict greater awareness of "ideal" survey responses among respondents (see Rossiter & Robertson, 1975, for a discussion of the "social desirability bias" in parents' self-reports). Indeed, one graduate-student mother responded to my inquiry about her attitude toward toy-related programming by stating: "I know the liberal answer is to object to these kinds of shows, but I really don't see anything wrong with my son watching them."

VCR USE COMPARISONS

Library owners are more active tapers, being significantly more likely to use their VCRs for time-shifting ($\chi^2 = 9.16, p<.01$) and (not surprisingly) for taping programs from the television set to keep in a library for their children ($\chi^2 = 22.3, p<.001$). Indeed, many respondents indicated that most or all of the videotapes they own were bought blank and are used for taping movies or other programs from the television. This pattern could reflect the parents' overall satisfaction with television programming as well as the cost-effectiveness of off-air taping. Because most of the children's fare available in video stores resembles that broadcast on television (see Wartella et al., 1990, for a description of the tapes available in the community), it makes economic sense for parents to compile their own collection of those television programs they deem acceptable for their children. And, because children often enjoy watching the same program several times, a small collection of "TV reruns" can provide an economical source of entertainment for children (see Dobrow, chapter 10, this volume).

VARIETY IN HOME VIDEOCASSETTE LIBRARIES

The final statistical analyses conducted address the question of videocassette library content—do parents compile a wide variety of tapes for their children or do they collect content from a limited number of genres? To explore the relationship between number and variety of tapes owned, a scale was again computed. Parents were not asked to list titles of their videocassettes, but were read a list of genres provided by local video store owners and asked to indicate which types of tapes they owned. Thus, a variety score was computed for each family by summing the total number of genres owned.

Library owners on the whole had a significantly more diverse selection of tapes than nonlibrary owners ($t = 7.94$, $p<.001$) and they also provided significantly more variety for their children ($t = 6.24, p<.001$).

It appears, then, that this sample of families does not generally use their VCRs for narrowcasting. However, there are always exceptions to the rule and in this sample they include the mother who reported that their entire video library (12 tapes) consisted of "Sesame Street" tapes for her 3-year-old daughter; the physician's wife who reported that all but one of their tapes show instructional surgical procedures; or the minister's wife who claimed all of their cassettes were biblical tales. So, although we can see an overall pattern of increased broadcasting with the VCR, a small minority of our sample uses their VCRs for narrowcasting purposes.

CONCLUSIONS AND DISCUSSION

This study attempted to assess whether the videocassette libraries compiled by families in one midwestern community serve to increase or decrease the variety of audiovisual content viewed by children. Earlier research suggested (Levy, 1980a) and discovered (Dobrow, 1989) that VCRs can fulfill a narrowcasting function for some users, allowing them to view content not available on commercial television. For these users, the VCR can provide alternatives to the hegemonic, or "mainstream" ideology consistently discovered in American commercial television fare (see Gitlin, 1987). Although this function has been identified in ethnic subcultures and groups opposing the dominant political order, it has not been examined in middle-class American samples. In this analysis, then, I examined the relationship between videocassette ownership and parental attitudes toward commercial broadcast and cable television. I hypothesized that parents who were not satisfied with the content of broadcast and cable programs for their children would use their VCRs to provide alternative, acceptable content.

The results indicate, however, that most parents use their VCRs as supplements to, not replacements for, broadcast and cable television. Tape libraries compiled by families contained a wide variety of tape genres, most of which reflect the variety already available on the TV medium. There could be a couple of different reasons for this pattern of ownership. First of all, this could reflect an overall satisfaction with the quality of content available on broadcast and cable TV, thus reducing the felt need for alternatives. Or, this ownership pattern could indicate that the inventory of cassettes available through video retailers does not provide many alternatives to what is available on television.

Both of these conditions were discovered in the Champaign–Urbana community. Parents reported moderate levels of satisfaction with commercial broadcast programming and higher levels of satisfaction for both public television and cable programming for children. Thus, it would appear that, in general, this group of parents did not feel the necessity to provide alternatives to televised content. However, an earlier examination of the videocassettes available through the retailers in the community (see Wartella et al., 1990) indicated that very few alternatives existed in the range of content available for children. Most of the children's tapes provided by local retailers resembled commercial television fare, with a majority being toy-related programs. So, even if parents were dissatisfied with the quality of children's television fare, their choice of video alternatives in the community is extremely limited.

There were indications in the data that some parents do limit the range of content available to their children; however, these parents are the exceptions, not the rule. The video libraries of younger children are more limited in number and scope than those of older children. This could reflect parents' greater degree of content control with younger viewers; or, it could simply reflect the differences in viewing appetites between older and younger children—younger viewers enjoy multiple viewings of the same program, whereas older children begin to demand less repetition. Thus, although a collection of one or two tapes may be sufficient to satisfy a preschooler, older children are more likely to want a wider selection. However, a wider selection does not necessarily require increased variety. It would be possible, for example, for parents to limit their children's viewing to the toy-related programming genre and still provide a large collection of videocassettes.

The case of the minister's family whose library was limited to biblical tales illustrates the opportunity for narrowcasting provided by VCR technology. The VCR allows viewers to replace mainstream television and cable fare with content more consistent with their values and ideologies, if such content is available.[2] Although this was not a very frequently reported use of the VCR among the families in this sample, further research with American subpopulations (i. e., different ethnic or religious populations) should explore whether these groups use their VCRs differently than this predominantly White, middle-class sample. Additionally, more extensive research with samples such as this one should attempt to uncover differences not located here. For example, is there a relationship between parents' choice of day-care programs for their children and their provision of videocassettes? Do parents who enroll their children in academically oriented programs differ from parents who enroll their children in programs that stress social development? If so, what are the

[2] See Ganley and Ganley (1987) for a discussion of the political uses of VCR technology to subvert government-controlled media in developing countries.

consequences of the viewing opportunities for children's development? Is there a difference by family structure in variety of tapes provided for children?

Videocassette libraries should be compared to print libraries for the content they provide and the functions they serve. Jordan's observational evidence, as discussed in this volume, found that children use videos as supplements to, not replacements for, books. Parents often offer children the option of a bedtime story or bedtime video, and the children in her sample were as likely to select a book as a video. But what about other samples of families? Do busy parents or parents with low reading ability replace the bedtime story with a bedtime video? If so, we must ask what effect, if any, this will have on children's development of reading skills. Salomon (1983) discovered that children often employ only basic decoding skills when watching television, but utilize more complex inference-making abilities when reading. The perceived "ease" of TV prompts less rigorous interpretational activities. He suggested that heavy viewers may become poor readers because of the lack of practice in abstracting concepts and making inferences. However, as Heath (1983) noted, reading alone does not stimulate higher levels of inference making. Parents must prompt their children to go beyond the presented information and encourage the application of printed information to their children's lives. Similar types of prompting have been shown to increase children's learning from TV (Salomon, 1977). Thus, parent–child interaction during video viewing and the uses to which videocassette content is put can influence children's abilities to perform more complex inferential processes during both print and audiovisual media consumption.

Dobrow (1989) reported that some of her respondents use their VCRs to teach their children about the homes and cultures left behind in their countries of origin. Drawing on the work of Gans (1974), she labeled these specialized videocassettes "subcultural programming," and suggested that this use of VCRs to facilitate ethnic identity may lead to the development of more pluralistic cultural perspectives among viewers. I propose that videocassettes can also be looked at according to Kellner's (1987) notion of "emancipatory popular culture." Not only can videocassettes fulfill the needs and wants of audience segments, but they can do it in a manner oppositional to the dominant social order. Without the requirement to attract a "mass" audience, videocassettes may provide ideological perspectives that are inconsistent with American "mainstream" media. Thus, parents who wish to expose their children to alternative ideologies may be able to do so through the use of VCR technology.

Foreign language videocassettes are not the only specialized genres available and further research should identify other such genres and audiences. We must ask what effect specialization and audience segmentation will have on the nature of commercial broadcast and cable television as well as on the concepts of a "mass audience" or "mainstreaming" (see Gerbner, Gross, Morgan, & Signorielli, 1980). Indeed, it appears that market specialization will not be limited to foreign language videocassettes, as industry analysts predict profits for sales of videocassettes into the next decade will come more from specialty cassettes than from theatrical movies (Mayer, 1988).

Examinations of videocassette content must ask whether VCRs will undermine hegemonic ideology or whether specialized programming continues to reinforce hegemony. Is there real choice or only an illusion of choice? Or, in the context of the variety concept used here, does owning multiple genres of programming actually increase the range of ideologies viewed? Do video alternatives to television and cable really offer opposition? Content analyses of different genres should address this question of difference.

In this analysis, I avoided discussion of the quality of the content of home videocassette libraries, choosing instead to focus on the question of variety. Quality, however, is certainly an important concept to consider when examining videocassettes available to children. Most academics would agree that a library limited to educational material is more beneficial to a child than one limited to toy-related programs. But because academic judgments do not always correspond with user's judgments, it is necessary to look further into parents' notions of quality and their evaluations of the kinds of content they provide for their children. Do parents look for videos with an academic orientation, a social orientation, or some other focus? What do parents consider "quality" videos?

Moreover, researchers should develop more fully theories of the social contexts of VCR use. This study was able to describe the amount and types of videocassettes available to children, but could not evaluate the influence of the content and viewing situation on the child's understanding or social interactions. Further research must extend these findings to examine how and why children watch videocassettes and to what purposes is the information gained applied. Rubin and Bantz (1987) found that their sample of adolescents and adults reported frequently using their VCRs for socializing purposes. Similarly, Lull (1982) described family television viewing as a time for substantial interpersonal contact. Do these styles of viewing extend to children's VCR viewing as well? What is the context of viewing for children and how does it influence VCR use?

The issue of parental control over children's media use has long been important to mass media researchers (Blosser & Heintz, 1987). In this study, parents reported no distinction between broadcast or cable television or VCRs when setting viewing guidelines for their children. This may be because parents do not recognize a difference between the content on the three providers. However, as was suggested earlier, this may be a function of the survey method used; parents may not be consciously aware of their behaviors toward the different media and assume consistency, or perhaps they purposely reported the perceived "desirable" higher levels of control for all media. Jordan's (chapter 9, this volume) observational evidence with a similar type of sample indicates that parents are less strict regarding VCR use than either broadcast or cable TV use.

This contrast identifies the need to validate survey data with observational evidence if possible. Due to the nature of the data gathered and the desire to sample widely in one community, observational methods could not have been easily managed for this study. Nevertheless, further research on VCR use should employ observational or ethnographic methods in order to understand more fully the complex relationship between media use and other lifestyle processes.

Finally, videocassette libraries should be examined in the context of the total media environment. For the families in this sample, videocassettes provide only one of many sources of mediated communication. The provision of a video library was not significantly correlated with children's time spent with the VCR, indicating that the presence of videocassettes did not lead to heavy viewing. Instead, I propose that the availability of a videocassette library may influence children's orientation to the medium. As Lin suggests (chapter 4, this volume) rather than a reactive, "receiver" orientation to the medium, the VCR owner may develop a more proactive, "user" orientation.

In conclusion, this study provides some valuable information regarding home video libraries, but when placed in the larger context of inquiry into a new communications phenomenon, illustrates the need for further, microlevel research to extend our understanding of the role of this medium in viewers' lives.

ACKNOWLEDGMENTS

Special thanks to Ellen Wartella, Amy Aidman, and Renzo Bustamante for all their efforts in making this chapter possible.

REFERENCES

Agostino, D., Terry, H., & Johnson, R. (1980). Home video recorders: Rights and ratings. *Journal of Communication, 30*, 28–35.

Barcus, F. E., & Hughes, C. (1988). *Local outlets for children's video in greater Boston.* Unpublished manuscript.

Blosser, B. J., & Heintz, K. E. (1987). *Rules for television viewing: An examination of ethnic differences and parent-child consensus.* Unpublished manuscript.

Dobrow, J. R. (1989). Away from the mainstream: VCRs and ethnic identity. In M. R. Levy (Ed.), *The VCR age* (pp. 193–208). Newbury Park, CA: Sage.

Dorr, A. (1986). *Television and children: A special medium for a special audience.* Beverly Hills, CA: Sage.

Federal Communications Commission. (1983). Report and order. *Federal communications commission reports, 96* (2nd series), 634–658.

Ganley, G. D., & Ganley, O. H. (1987). *Global political fallout: The VCR's first decade.* Norwood, NJ: Ablex.

Gans, H. J. (1974). *Popular culture and high culture.* New York: Basic Books.

Gerbner, G., Gross, L., Morgan, M., & Signorielli, N. (1980). The "mainstreaming" of America: Violence profile no. 11. *Journal of Communication, 32,* 10–29.

Gitlin, T. (1987). Prime time ideology: The hegemonic process in television entertainment. In H. Newcomb (Ed.), *Television: The critical view* (4th ed., pp. 507–532). New York: Oxford.

Greenberg, B. S., & Heeter, C. (1987). VCRs and young people. *American Behavioral Scientist, 30,* 509–521.

Heath, S. B. (1983). What no bedtime story means: Narrative skills at home and school. *Language and Society, 11,* 49–76.

Kellner, D. (1987). TV, ideology, and emancipatory popular culture. In H. Newcomb (Ed.), *Television: The critical view* (4th ed., pp. 471–503). New York: Oxford.

Levy, M. R. (1980a). Home video recorders: a users survey. *Journal of Communication, 30,* 23–27.

Levy, M. R. (1980b). Program playback preferences in VCR households. *Journal of Broadcasting and Electronic Media, 24,* 327–336.

Levy, M. R. (1983). The time-shifting use of home video recorders. *Journal of Broadcasting and Electronic Media, 27,* 263–268.

Levy, M. R. (1987). Some problems of VCR research. *American Behavioral Scientist, 30,* 461–470.

Lull, J. (1982). The social uses of television. In G. Gumpert & R. Cathcart (Eds.), *Inter/media* (2nd ed., pp. 566–579). New York: Oxford.

Mayer, I. (1988, December). Goodbye easy growth. *Channels/Field Guide 1989,* p. 106.

Mayer, I., & Sweeting, P. (1986, December). The next push for growth. *Channels/Field Guide 1987,* p.92.

Mayer, I., & Sweeting, P. (1987, December). There's many a way to get that $2.25 a night. *Channels/Field Guide 1988,* p. 128.

Rossiter, J. R., & Robertson, T. S. (1975). Children's television viewing: An examination of parent-child consensus. *Sociometry, 38,* 12–25.

Rubin, A., & Bantz, C. R. (1987). Utility of videocassette recorders. *American Behavioral Scientist, 30,* 471–485.

Salomon, G. (1977). Effects of encouraging Israeli mothers to co-observe "Sesame Street" with their five-year-olds. *Child Development, 48,* 1146–1151.

Salomon, G. (1983). Television literacy and television vs. literacy. In R. W. Bailey & R. M. Forsheim (Eds.), *Literacy for life: The demand for reading and writing.* New York: The Modern Language Association.

Speculating on new media's effects. (1981, January 12). *Broadcasting*. p.74.
Wartella, E. A., Heintz, K. E., Aidman, A., & Mazzarella, S. (1990). Television and beyond: Children's video media in one community. *Communication Research, 17*, 45–64.
Zoglin, R. (1988, October 17) The big boys blues. *Time*, pp. 56–61.

9

A Family Systems Approach to the Use of the VCR in the Home

Amy B. Jordan
Widener University

Since the early 1980s, communications researchers have increasingly recognized the family as an important context for understanding media behavior. When television first became widespread, there was much concern over the impact of the medium on family life (Maccoby, 1954; McDonagh, 1950). Researchers warned that television was stealing precious moments of interaction time between husband and wife and between parents and children. This interest in the family, however, was short-lived. As the public and press became more comfortable with the presence of the electronic stranger in the home, attention shifted to the "deleterious effects" of television content on susceptible individuals (primarily children) and on more narrow and multi-contextual topics (for example, the effect of violent television on aggressive behavior). The family as a context and mediating variable for individual exposure to television was all but forgotten.

Lull's (1980) research on the social uses of television within the home brought the family context back into focus. Working from the uses-and-gratifications tradition, his studies had an interesting twist. Instead of examining what television does to families, he explored what the family does to television. This perspective revealed that families are active, conscious, and purposeful in the ways they use the medium.

Obviously, most media are located and utilized within the confines of the home. Nearly every American family owns at least one television set and at least one radio. Moreover, half of all families subscribe to newspapers and magazines (Gollin & Anderson, 1980). New technologies such as the VCR are entering the home at a surprising rate. The home environment, it has been argued, is rapidly becoming a *media-rich* environment (Bachen & Jordan, in press).

Mass media have become inextricably woven into nearly every aspect of family life. The rhythm of family time is often set by the patterns of media use (Bryce, 1987; Leichter et al., 1985; Lull, 1980). For example, mornings may be initiated by newspaper reading, evenings may center on television programs, and bedtimes may be structured around story reading. Interactions between family members are shaped in and around media content (Lull, 1980; Morley, 1988), and media settings offer a stage for the playing out of gender roles (Bryce & Leichter, 1983; Morley, 1988). Even space within the home is structured by media. For many family living rooms, the television set is the dominant piece of furniture and the clear center of attention (Bryce, 1987; Lindlof, Shatzer, & Wilkinson, 1988). This chapter, therefore, argues that if we are to understand media use by individuals, we first need to understand media use by families.

To some extent, research on the VCR has incorporated family demographics or other elements of family life in order to understand individual's behavior with this new technology. Early researchers attempted to understand why some families got a VCR, whereas others did not, and often turned to family or home characteristics (such as social class, possession of other media technologies, age of family members, etc.) to explain the phenomenon (Greenberg & Heeter, 1987). Klopfenstein (1988) has found, for example, that homes with children are more likely to have VCRs than families without children. Other research has examined what people watch on the VCR and how viewing varies according to the sex and age of the individual. Researchers have also looked at what happens to the overall use of media when the VCR is introduced to the media package already available to family members (Hughes & Dobrow, 1988; Levy, 1987).

Very little research, however, has been concerned with how VCR behavior has emerged from or contributed to the overall patterns of media or other behaviors within the home, although there are a few exceptions. Kim, Baran, and Massey (1988) examined the ways in which VCR use affects parental control of children's television viewing. Dobrow (1988) has explored the use of the VCR in facilitating ethnic identity in small groups, some of them families. Lindlof and colleagues (1988) examined the differential use of the VCR by family members and its relationship to roles within the home. And so-called "social uses" of the VCR by families and friends were explored by Gunter and Levy (1987).[1]

---

[1] They are "so-called" social uses because Gunter and Levy (1987) found that most VCR viewing that occurs within the home is solo viewing.

Although these studies go beyond merely describing VCR behavior and come closer to explaining the patterns, there is still much work to be done in order to understand the origins and maintenance of such patterns of VCR behavior. Clearly, we need to understand VCR use as a *part* (albeit a significant part) of other media use within the home. Moreover, it is necessary to understand behavior around the VCR in the context of other patterns of family behavior. For example, Lin and Atkin's (1988) study on parental mediation of television and VCRs revealed a positive correlation between rules for television and rules for the VCR. Why is this the case? Are rules surrounding the media reflective of a pervasive system of rule-making in the family? Although it is certainly important to uncover who uses the VCR for what purposes, this chapter presents research that goes a bit deeper in order to find the connections between VCR use in the home and the system or norms, values, and beliefs of the family. Moreover, an attempt is made to place the research within a larger theoretical framework.

BUILDING A THEORETICAL FRAMEWORK

One theory that has recently been employed by communications researchers to place media use in the context of the family environment is the Family Systems Theory (Bachen & Jordan, in press; Goodman, 1983).[2] Family systems theory grew out of the broader General Systems Theory, whose basic premise is that one must study *wholes* rather than *parts* if one is to understand how the system functions (Bavelas & Segal, 1982; Vetere & Gale, 1987). The General Systems Theory was adapted and implemented by family therapists who conceptualized individuals as members of an ecological system.[3] They considered how the distinct aspects of the family work to establish and regulate the functioning of the *whole* family despite internal and external pressures to change. For example, researchers note that time-shifting (recording programs for playback at a more convenient time) is one common use of the VCR in the home. Family systems theorists might ask: Does the use of the VCR for time-shifting reflect other values the family has about time? Such

[2] Rogge and Jensen (1988) and Lindlof et al. (1988) described the family as a "system" but do not frame their work in Family Systems Theory, per se.

[3] In the 1950s, clinicians in psychology began to apply systems theory concepts to their patient analysis. Most notable were Bateson and his colleagues, who extended the theory to schizophrenic symptoms. With the systems approach, a schizophrenic's behavior was viewed as an adaptive response to a dysfunctional family environment (Bavelas & Segal, 1982).

questions allow us to see the VCR as one component of the home that both reflects and shapes the overall system of family life. This approach represents "an epistomological turn away from thinking of 'forces' and 'causes' toward thinking of 'relationships' and 'contexts'; . . . away from linear models towards recursive or circular descriptions" (Bochner & Eisenberg, 1987, p. 4).

Research on the role of the VCR in family life might benefit from a grounding in Family Systems Theory. Studies on mass media use have traditionally looked at the individual and the media in isolation from the family context, and studies that focus on the VCR are no exception. Typically, surveys and interviews have been employed in VCR research. Although such methodologies have been important in providing an overall picture of who uses the VCR for what purposes, they fail to get at some of the complex issues of social activity and collective meaning making around the medium.

AN ETHNOGRAPHNIC EXAMINATION OF VCR USE IN THE HOME

The work described in this chapter is part of a larger study that examines the role of the media in family life. Specifically, the study asks: How does mass media use emerge from and contribute to both the component parts and the overall structure of family life? This chapter concentrates on VCR use in the family system, and the ways in which behaviors around the medium are reflective of the norms, values, and beliefs of the family.

Family systems theorists argue that to understand the system, one can neither remove it from the natural setting (the home) nor study family process outside of its original temporal occurrence. The unit of analysis, moreover, must be the family as a *whole*. Within this whole, subsystems are examined for their relationship to each other and to the larger system.

To explore this area through a naturalistic approach, I employed an ethnographic method. Ethnographic methodologies are rarely used in mass media research, but a handful of studies provide examples of what can be accomplished through participant observation and in-depth interviews (see Bryce & Leichter, 1983; Lindlof et al., 1988; Lull, 1980; Palmer, 1986; among others).

On the basis of past research, I decided that a 3-day observational period with a follow-up interview at the end of my stay would provide the depth and variety of information needed to explore the role of the media in the family system. I spent approximately 1 hour with the family in the morning (until they went off to work and school) and 3 to 4 hours in the evening (until the children went to bed) for an average of about 12 observational hours per family. Data were obtained from tape recordings of family interactions, written notes I took during and after the observational period, and the interviews I conducted with family members.

Twenty-one families from the suburbs of a Northeastern city were observed over the course of 11 months in 1988 and 1989. The families were unknown to me and were recruited from churches, schools, and friends of friends. Because the sample is small, I controlled for a number of variables in order to make the families comparable. White, two-parent families, with two or three children under 12 years of age were selected. Moreover, in all families both parents had paid employment. The sample consisted of an equal number of working-class, middle-class, and upper middle-class families.

This technique is not one without problems. Families may change their behavior in response to this "outside stranger." I believe, however, that working with families over a 3-day period prevented sustained dramatic changes. Parents were asked at the end of the observational period what difference my presence made in their daily patterns and typical interactions. Invariably, parents said that there were few or no significant changes, (although a couple of parents did point out that the family tended to be more fully clothed on the mornings of the study period). Moreover, the children tended to keep things "honest"—they would complain or comment on changes in parental behavior. In one family the mother kept turning off the television set when it was not being watched. Each time she did this, the children became agitated. Finally, the 9-year-old boy said, "Mom, quit it. You never do that when Amy isn't around."

SOME INITIAL FINDINGS

The data from the observations and interviews are organized around a set of family domains that appear to be illustrative of family life and the media's role within the family system. These domains include the *structural domain* (time and space within the home), and the *social domain* (roles, rules, and interactions in the home). Clearly, these domains are related to each other and, in family systems terms, to the overall goals of the family. Also, these are not the only domains of family life. However, previous research as well as exploratory fieldwork uncovered these areas as fruitful realms of study.[4] The following is a description of the role of

[4] Lull's (1980) "typology" of social uses of television has provided the basis for the description of "domains" in the present study.

the VCR in 15 families, along with related research in the areas of family, media, and VCR use in the home.

Structural Uses of the VCR in the Home

Space and time in the home are two elements of family life rarely considered in studies of mass media use. Although television's influence on family interactions is important, how media fit into the time and space in the home may shed new light on family's construction of a media environment. Moreover, focusing on the role of the media in family life brings us closer to characterizing the relationships of the structural and social domains to each other and to media usage in the home.

The VCR and Spatial Structuring. The importance of space to human interaction and functioning has long been recognized by researchers in the social sciences. Most predominant has been research by Hall (1966), who argued that spatial environments have both norms and implications. Space is important in the home, as it regulates relationships, activities, and the private and public behavior of individuals within the family. The home has its own unique space. Rooms are organized around functional principles (kitchens, bedrooms, living rooms) that serve to draw people together or keep them apart (Ashcraft & Scheflin, 1976). How the VCR fits into these "spaces" may be an indication of how the VCR fits into the larger system of family life.

One could argue that the prevalence of a medium in a room is related to the importance of that medium to the family. Television, for example, occupies a central position in many gathering areas. Families often structure their living areas around the television set—arranging furniture for ease in viewing rather than conversation (Leichter et al., 1985; Lindlof et al., 1988; Morley, 1988). In my study, like others (Lindlof et al., 1988), affluent families had many television sets throughout the home, encouraging separate or private viewing rather than family viewing. In the case of the VCR, however, there was often only one machine and it was located in the main gathering area with the biggest and best TV. In all but two families, VCR use occurred in the den or living room area of the house.[5]

The viewing of videocassette tapes is generally less spontaneous than the viewing of television. Planned viewing allows the family the opportu-

[5] The families had VCRs in the parents' bedrooms as well as the living rooms. One especially affluent family had a VCR in the car for the children to watch on long trips.

nity to "get comfortable." In this research, I found a family's process of getting ready for viewing often required a rearranging of space. People moved chairs closer to the TV, children got blankets, toys, bottles and snacks, and lights were dimmed (if it was evening). Families settled in when viewing material on the VCR to a much greater extent than they did with television viewing—partly because it was a more purposeful activity, and partly because the running of the tape could wait while everyone got ready.

VCR space might also be thought of in another way too—that is, the space a VCR library takes up. VCR libraries have been discussed elsewhere in this book (chapter 8). In some ways, the VCR library may be like the print library. For example, families that display large libraries of books may be emphasizing the importance of print media in their home. Similarly, families with many videotapes prominantly displayed may be seen as valuing the VCR as a resource within the home. In fact, families in my study with videotape libraries often made a space for the tapes in bookcases right along side the books. This illustrates Levy and Fink's (1984) notion that tapes become competition for other "on-shelf" or "on-demand" items such as the print media. Like books, videotapes can be used repetitively for entertainment and information.

On-demand uses of the VCR emerge as an important function in the present study. Videotape viewing often replaces book reading as a bedtime ritual for children (as is discussed in greater detail later in this chapter). Moreover, videotapes are a resource for school-aged children working on educational projects. For example, one mother told me that her son had to write a report about space exploration and that he had been working "with a bunch of useless books." He was almost finished when she remembered that they had taped a TV series on space. She said: "Now that's a functional use of media. If there were a broader range of videotape information stuff, then we'd start collecting that."

VCRs and Time Use in the Home. The time-use ideology of families has been increasingly recognized by communication scholars—particularly those interested in how families use television (Bryce, 1987; Bryce & Leichter, 1983; Medrich, Roizen, Rubin, & Buckley, 1982). How families think about time—for example, time as a scarce resource that needs to be budgeted and maximally utilized versus time as plentiful and something that can be wasted—may indeed have an impact on how television and other media fit in with the temporal rhythm of the home. The VCR is an especially interesting medium because with this new technology television loses what Levy and Fink (1984) called its "temporal transience." Programs can be taped for later viewing, movies can be rented to fill a particular time slot (e. g., weekend nights), and network commercials can be fast-forwarded, diminishing the amount of time required to watch a show (see Massey & Baran, chapter 5, this volume).

In my research, families' use of the VCR to control media time can be classified in four general ways: (a) to *shift* when a program is viewed, (b) to *adapt* how much and what portions of a tape are watched, (c) to *structure* time for the family (particularly bedtime), and (d) to *fill* time that seems empty. Moreover, families' use of the VCR to shift, adapt, structure, or fill time is consistent with the ideology of time within the home.

Time-shifting is a function of the VCR that allows a program to be taped and saved for later playback (Levy, 1983). The actual extent to which time-shifting occurs is unclear. Levy found that although many VCR owners tape programs for later viewing, almost half of the tapes never get played back. A similar phenomenon occurred in my study. Parents who taped shows often never "got around" to viewing them. Unless the show was a favorite and there was a strong commitment to watch it, the tape would go unwatched and reused for another program taping. One 12-year-old, in talking about his parents' VCR use, said "they tape all these shows off TV and then never watch them." Mother: "Yes we do!", then, "Well, every once in awhile we watch them."

Parents in this study were more likely to use the VCR to time-shift for their children. Shows that were recorded tended to be ones that were on after their children's bedtime (e. g., PBS's "Nova") or programs that might have been missed due to other activities. Two families taped up to 4 hours of Saturday morning cartoons, because weekends were often the only time these busy parents could spend with their children. The cartoons were then viewed throughout the week. Parents also encouraged the use of the VCR to tape favorite prime-time programs if homework was not done. One mother told me:

> If Daniel watches TV and then comes in and tells me he has too much homework then he gets this nasty lecture about "how many times have I told you that if you haven't finished your homework then tape the damn thing and watch it later." Because we have two videotape [machines] and there's no reason in the world why he can't tape it.[6]

In my conversations with the children, however, I discovered that taping was not something they chose independently. It became apparent that watching TV can be a technique to *avoid* doing homework (or other activities). Applying this logic to research on time-shifting helps to explain why more time-shifting does not occur. The fact that time-shifting TV programs is more the exception than the rule is often surprising to

[6] Names here and throughout the chapter have been changed to protect the identity of the families.

scholars who believe that watching during convenient times and fast-forwarding through commercials is preferable to viewing in "real time." Consciously or unconsciously, the families I worked with seemed to enjoy the slower, more structured "pace" of real time viewing and may prefer watching no TV to watching taped TV. In fact, many people in this study had never bothered to learn how to use the automatic timer on the VCR to record a program when the family would be away from the home: "We don't tape as much [*sic*] shows as we should. See, we never figured out how to set the timer on the VCR. I don't know why." Other studies have found a similar ignorance of the VCR features (Lindlof et al., 1988).

A second way in which the VCR plays a role in family time is its ability to adapt time. VCRs allow individuals to fast forward past commercials, view portions of a program repeatedly or not at all, and pause a program in order to pursue another activity briefly. One father also told me that when he tapes programs for his children he pauses the tape during commercial breaks, effectively using the VCR to edit what his children will view. These adaptative time-use functions have emerged frequently in studies with adults (Levy & Fink, 1984) and adolescents (Hughes & Dobrow, 1988).

In this study, parents used the fast forward and pause function of the VCR frequently. Children, on the other hand, watched a VCR tape in much the same way they would watch a broadcast TV program—from beginning to end with no manipulation of time. One mother pointed out that her 6-year-old child did not really understand the options available on the VCR, or, if she did, she had not figured out how to work them properly. Another mother told me that her children *like* to watch commercials. When she fast forwards past commercials, they become upset.

A third relationship between families' VCR use and time in the home is the role of the VCR in the structuring of the day. Lull (1980) and Leichter et al. (1985) have found that television's schedule serves to punctuate bedtime, dinnertime, bathtime, and so on. VCRs have a similar function. Although tapes do not necessarily start on the hour or the half hour, they do fit in to clear temporal boundaries. For example, many parents used the VCR to structure "quiet time" within the home. (Quiet time is usually the time in between school and dinner when the children wind down from their busy day.) The VCR was also used as part of the bedtime ritual, with bedtime commencing as soon as the tape was over. Interestingly, children were often offered a choice: They could watch a videotape or read a book. Most often, the tape won out, although sometimes the children were allowed to do both. If this was the case, however, strict time limits were generally set.

> Mother: Do you want to watch Dumbo? You can watch the whole thing.

[The 3-year-old girl says yes.]

Mother: If we hurry, then we have just enough time to watch all of Dumbo tonight. But then, no story. We can't do Dumbo and a story. You can watch all of Dumbo and no story or half of Dumbo and a story.

Rita: All Dumbo, no story.

Mother: OK, let's go watch all of Dumbo.

Finally, the VCR can be used to fill time for both children and parents. Because I worked exclusively with families where both parents had jobs, one of the most common uses of the VCR was as "babysitter." Although this function has been recognized for television use (Gantz & Masland, 1986), my research indicates that it may also be prevalent with the VCR. In fact, I witnessed that tapes were often more reliable than broadcast television in engaging children's attention. Also, parents who felt uncomfortable about spending time away from children felt less guilty and more in control with a tape than with on-air TV because they knew what they were watching.

> Most of the time, the kids watch tapes . . . I don't mind that its routine. It's a good babysitter in the morning, which we need the time to get ready for work and it's not on that long in the morning. It's nothing worse than they see on television.

VCR Use in the Family's Social Domain

Lull's (1980) "social uses" typology has contributed to current thinking about VCR use in the social domain of family life. Although he concentrated solely on television use, Lull's attention to the media's role in family interactions and relationships has been important in advancing our understanding of the meaning of media to individuals and families. My research turns to VCR use by families, and redefines the social domain as the interconnection of roles, rules, and interactions that make up the family system.

VCR Use and Family Roles. Individuals in families adopt roles that are played out in various ways within the home. For example, mothers may be more supportive and fathers may be more directive. Older siblings may tell younger siblings how to behave and what to do. Although it may be frustrating for those who feel uncomfortable with their ascribed roles, role-taking and role-playing aid family functioning by "providing a means for the reciprocal prediction of behavior" (Brody & Stoneman, 1983, p. 339).

Family roles are manifested in media contexts. VCR use is a particularly interesting stage for the display of role behavior as there are selections to be made, machine operations to be assumed, and remote controls to be managed. In this section, I concentrate on the gender-related behaviors of parents and role interactions of siblings, although a host of other role behaviors occur around VCR use.

Lull (1982) and Morley (1986, 1988) have found that fathers generally control program decisions and remote boxes in the television viewing situations. This may be true for television, but my research indicates that mothers controlled and coordinated VCR use. When family programs were taped off-air or movies were rented, mothers were typically the ones who thought ahead and carried out the task. The mother would often tell the family what she was doing so that they were aware of their choices. When movies were rented for parental viewing, husbands sometimes had veto power over selections (although the mothers went to the store to get the tape).

Little research has been conducted on sibling roles within media situations. Zahn and Baran (1984) asked college students to recall "who typically determined what is to be viewed?" Not surprisingly, those students who were youngest siblings said they won program choice least often. My research indicates a similar pattern. Older siblings controlled the physical manipulation of the VCR (and their mechanical skills with the medium were a source of pride). They were more likely to determine the choice of the program as well. I witnessed few conflicts over tape selection, perhaps because a videotaped program can be seen at a later time if not viewed in the present. Moreover, younger siblings tended to concur with their older sister or brother's selection. One expects this to be the case—an older sibling's dominant position in the family is easily translated into a variety of domains, VCR use being but one.

Family Rules and Control of VCR Use. Medrich et al. (1982) found that families with two working parents typically impose fewer rules on television viewing, and my results are consistent with that finding. Few rules existed surrounding VCR use in the families observed for this study. There were also few rules regarding media use by children in general. The mothers in this sample were, however, quite familiar with the material on tape in the video libraries. Because mothers were acquainted with the choices, there was a greater feeling of confidence that they knew what their children were seeing and approved of it.

VCRs were used by many mothers to manage unacceptable on-air network television. A number of times, I witnessed a mother disapprove of a program her son or daughter was watching and suggest a videotape instead. This was a more successful, and less conflictual alternative to turning the television set off. One mother came home from work to find her babysitter and daughter watching a shoot-out on TV.

Mother: What is this? What is this nonsense on TV?

Babysitter: Disney.

Mother: Disney? There's got to be something less violent. Rita, do you want to watch Dumbo? That's less violent.

Rita: I want to watch Dumbo.

Parental control over VCR use seems to reflect an overall picture of time, rules, and media use by children in the home. Many families had rules regarding bedtime and attitudes regarding the flexibility of bedtime. If bedtime was strict, the VCR was used by families to tape programs that were on after the children's bedtime (thereby reinforcing the pre-existing time structure and avoiding conflict over program viewing). If bedtime was flexible, a post-bedtime show that children wanted to watch may not have been taped. Instead, children were allowed to stay up late. The viewing of television in this case because a special event for the children.

The Role of the VCR in Family Interaction Patterns. There is little scholarly agreement on the nature of family interaction around VCR viewing in the home. Although Gunter and Levy (1987) said that VCR use is a "privatized media experience, often unshared between members of the same household" (p.491), Einsiedel and Savage's (1988) survey indicates that some 50% of viewing is done in the company of family members. My research leads me to believe that one cannot characterize VCR use in families with young children in a simple way. Multiple uses of the medium create a variety of viewing situations in the home. These viewing situations, moreover, appear to reflect the multidimensional needs of the family members for psychological affiliation, physical closeness, and personal space.

Children's tapes were generally viewed by children alone. Parents would set up tapes for their children to watch and would often proceed to engage in other activities in different rooms. For parents, video viewing was a means to distract and entertain the children in order to obtain some time or space to get their own tasks accomplished. In interviews, parents often called this the "babysitting function" of the VCR.

The viewing of children's videotapes, on the other hand, gave rise to extensive interaction between siblings who viewed together. This was particularly true if the tape was often viewed and well known. Although the older sibling was more attentive to the screen than the younger, he or she was often willing to engage in conversation or play initiated by his or her younger brother or sister. The tape became a background to a new primary activity, and the children selectively attended to the screen for the "good parts" of the program or movie.

Another VCR viewing scenario occurred with family viewing. Tapes watched by the family as a unit tended to be rented movies, often musicals, or favorite programs previously recorded off network TV. Unlike television viewing, families showed a greater interest in the program material. They took fewer breaks in viewing because commercials, if present, were fast-forwarded. This led to little verbal interaction during the programs by family members.

Families who viewed a tape together, on the other hand, would talk about what they had seen together in contexts outside viewing. Movie plots or other program content became fodder for conversations at the dinner table. Parents also used shared viewing to educate their children. For example, one mother tried to teach the concept of size to her 3-year-old daughter through a discussion of *Alice in Wonderland*.

Mother: "When [Alice] drank the first bottle, what happened? She got very . . ."

Cassie: "Big!"

Mother: "Right. And when she drank the second bottle, what happened? She got very . . ."

Cassie: "Little!"

Mother: "Right. Did you know that big and little are opposites?"

Another VCR use in family affiliation is parental viewing of rented movies on weekend nights. During my interviews with parents, mothers and fathers frequently told me that this activity was one of the few things they did in the home together without their children. This kind of viewing allowed for little verbal interaction, but provided rare time alone for the parents to reconnect. Busy parents with little time for themselves and for each other welcomed the presence of the VCR in the home as a means of creating a "date-like" situation for them. In fact, many parents labeled weekend rented movies "date tapes."

DISCUSSION AND CONCLUSIONS

Previous research on the videocassette recorder has mainly focused on two things: who gets a VCR and what the medium is used for. This chapter has attempted to move in the direction of understanding the role it plays in the context within which most VCR use occurs—the family. To parcel out the impact of the VCR on family patterns or, on the other hand, the influence of family patterns on VCR behaviors would be fruitless. Instead, I argue that family behaviors and media characteristics are related to both each other and to the larger family system from which they emerge.

Observations and interview questions exploring the role of the VCR in the structural domain of the home uncovered that families have norms relating to time and space that are reflected in VCR behaviors. The placement of the VCR and its accessories, as well as the arrangement of furniture around them, give clues regarding the value of the medium and its role in creating communal gathering spaces. Moreover, as Leichter et al. (1985) pointed out: "the organization of space both reflects and influences the multiple spheres of interpersonal interaction that occurs within families" (p.30). Thus, the finding that most living rooms are arranged to facilitate viewing rather than conversation parallels the family norm that discourages talking during VCR use.

Beliefs about the value and management of time are also reflected in the family's use of the VCR. As working parents, both mothers and fathers have obligations that take them away from their children. The VCR's ability to fill in empty or alone time for children emerges as a valuable resource for parents. Moreover, working parents with young children view time as a precious commodity. From this standpoint, the family's use of the VCR to shift, adapt, and structure time is a reflection of parental need to control the temporal environment of the home.

An examination of the role of the VCR in the social domain also illustrates the notion that VCR patterns emerge from the larger patterns of the family system. The context of VCR viewing offers a stage on which gender and age-related roles of family members can be articulated. Mothers in this study typically assumed control of the taping of programs and the renting and previewing of videocassettes. This behavior is a reflection of their overall role as primary caretaker of the children and the house. Similarly, older siblings dictated program selection and VCR operation more often than younger siblings. It seems natural that such patterns would spring from the dominant position he or she normally assumes in non-media contexts.

In this study, families had few strict and explicit rules regrading VCR use. A basic question, however, needs to be considered: How do we define control? Control can be seen as the rules about what can be viewed by children, which family members are allowed to operate the machine, and how much time can be spent with the medium. Or, control can be found in the form of positive reinforcement—that is, the suggestion of programs to be taped, movies to be rented, or on-shelf tapes to be viewed. My research uncovered instances of each form of control. Parents negotiated VCR use with children in such a way that it fit with other family rules regrading bedtime, homework, and parental notions of acceptable video content.

The VCR plays a visible role in the interaction patterns of the family. Although conversation is not typical during viewing, I have found that VCR content finds its way into conversations outside of the viewing context. Shared viewing allows for a common agenda between family members. Moreover, rented movies play an important role in parents' efforts to find space and time for themselves in a home where privacy is the exception rather than the rule.

Thus far, I have discussed the structural and social domains of family life separately. If, however, we are to hold true to the tenets of Family Systems Theory, we need to recognize that each domain is connected to the other and to the overall system of norms and values in the home. Parental collection of children's videotapes, for example, may be the integration of family beliefs about the management of time and the control of media content. Further, norms within the family are often gleaned from belief systems external to the home. The use of videotapes to structure bedtimes may be seen as a continuation of the oral tradition that teaches children norms and values through stories.

In summary, this research illustrates the notion that there is a reciprocal and symbiotic relationship between media behavior and family life. By tying individual uses of the VCR to the larger system of the family, we are able to come to a deeper understanding of the patterns of VCR use that have emerged within the home.

ACKNOWLEDGMENTS

Special thanks to Barbie Zelizer and Diane Umble for their comments on an earlier draft of this chapter.

REFERENCES

Ashcraft, N., & Scheflin, A. E. (1976). *People space: The making and breaking of human boundaries.* Garden City, NY: Doubleday.

Bachen, C. M., & Jordan, A. B. (in press). Mass media and academic achievement within the family system. In R. Pedone (Ed.), *Children, television and education.* Westport, CT: JAI.

Bavelas, J. B., & Segal, L. (1982). Family systems theory: Background and implications. *Journal of Communication, 32*(3), 99–107.

Bochner, A. P., & Eisenberg, E. M. (1987). Family process: System perspectives. In C. Berger & S. Chaffee (Eds.), *Handbook of communication science* (pp. 540–564). Newbury Park, CA: Sage.

Brody, G. H., & Stoneman, Z. (1983). The influence of television viewing on family interactions: A contextualist framework. *Journal of Family Issues, 4*(2), 329–348.

Bryce, J. W. (1987). Family time and television use. In T. R. Lindlof (Ed.), *Natural audiences: Qualitative research of media uses and effects* (pp. 121–138). Norwood, NJ: Ablex.

Bryce, J. W. & Leichter, H. J. (1983). The family and television: Forms of mediation. *Journal of Family Issues, 4*(2), 309–328.

Dobrow, J. R. (1988, May). *Away from the mainstream? The role of VCRs in facilitating ethnic identity*. Paper presented at the International Communications Association, New Orleans, LA.

Einsiedel, E. F., & Savage, D. (1988, May). *VCR usage patterns among a rental segment: Expanding definitions of new technology use in a Canadian sample*. Paper presented at the International Communications Association, New Orleans, LA.

Gantz, W., & Masland, J. (1986). Television as babysitter. *Journalism Quarterly, 63*, 530–536.

Gollin, A. E., & Anderson, T. (1980). *Mass media in the family setting: Social patterns in media availability and use by parents*. New York: Newspaper Advertising Bureau.

Goodman, I. (1983). Television's role in family interaction. *Journal of Family Issues, 4*(2), 405–424.

Greenberg, B. S., & Heeter, C. (1987). VCRs and young people: The picture at 39% penetration. *American Behavioral Scientist, 30*(5), 509–521.

Gunter, B. & Levy, M. R. (1987). Social contexts of video use. *American Behavioral Scientist, 30*(5), 486–494.

Hall, E. T. (1966). *The hidden dimension*. Garden City, NY: Anchor Press.

Hughes, C., & Dobrow, J. R. (1988, May). *The VCR and the adolescent: Patterns of use*. Paper presented at the International Communications Association, New Orleans, LA.

Kim, W. Y., Baran, S. J., & Massey, K. K. (1988, May). Impact of VCR on control of television viewing. *Journal of Broadcasting and Electronic Media, 32*(3), 351–358.

Klopfenstein, B. C. (1988, May). *The emerging VCR household: Relationships among ownership, demographics, and usage patterns*. Paper presented at the International Communications Association, New Orleans, LA.

Leichter, H. J., Ahmed, D., Barrios, L., Bryce, J., Larsen, E., & Moe, L. (1985). Family contexts of television. *ECTJ, 33*(1), 26–40.

Levy, M. R. (1983). The time-shifting use of home video recorders. *Journal of Broadcasting, 27*(3), 263–269.

Levy, M. R. (1987). Some problems of VCR research. *American Behavioral Scientist, 30*(5), 461–470.

Levy, M. R., & Fink, E. L. (1984). Home video recorders and the transience of television broadcasts. *Journal of Communication, 34*(2), 56–71.

Lin, C. A., & Atkin, D. J. (1988, May). *Parental mediation and adolescent uses of television and VCRs.* Paper presented at the International Communications Association, New Orleans, LA.

Lindlof, T. R., Shatzer, M. J., & Wilkinson, D. (1988). Accommodation of video and television in the American family. In J. Lull (Ed.), *World families watch television* (pp. 158–192). Newbury Park, CA: Sage.

Lull, J. (1980). The social uses of television. *Human Communication Research, 6*(3), 197–209.

Lull, J. (1982). How families select television programs: A mass observational study. *Journal of Braodcasting, 26*(4), 801–811.

Maccoby, E. E. (1954, Fall). Why do children watch television? *Public Opinion Quarterly,* pp. 239–244.

McDonagh, E. C. (1950). Television and the family. *Sociology and Sociology Research, 35,* 113–122.

Medrich, E. A., Roizen, J., Rubin, V., & Buckely, S. (1982). *The serious business of growing up.* Los Angeles, CA: University of California Press.

Morley, D. (1986). *Family television: Cultural power and domestic leisure.* London: Comedia.

Morley, D. (1988). Domestic relations: The framework of family viewing in Great Britain. In J. Lull (Ed.), *World families watch television* (pp. 22–49). Newbury Park, CA: Sage.

Palmer, P. (1986). *The lively audience: A study of children around the TV set.* Sydney, Australia: Allen & Unwin.

Rogge, J., & Jensen, K. (1988). Everyday life and television in West Germany: An empathic-interpretive perspective on the family as a system. In J. Lull (Ed.) *World families watch television* (pp. 80–115). Newbury Park, CA: Sage.

Vetere, A., & Gale, A. (Eds.). (1987). *Ecological studies of family life.* New York: Wiley.

Zahn, S. B., & Baran, S. J. (1984). It's all in the family: Siblings and program choice conflict. *Journalism Quarterly, 61,* 847–852.

10

The Rerun Ritual: Using VCRs to Re-View

Julia R. Dobrow
Boston University

By now, everyone knows that the VCR is a technology that allows for great flexibility in viewing. This is not only true in terms of content (enabling people to watch *what* they want to watch) and time (enabling people to watch *when* they want to), but also in terms of form. There already exists a small (but growing) literature on how and why VCR users "zip," or fast-forward through video content. This is a chapter on why they go backwards. Over, and over again.

The question of why people deliberately select to review content that they have already seen is a curious one. Why would anyone want to see a movie multiple times? Doesn't it spoil the experience if you already know the ending? And perhaps even more curious are those people who watch televised content more than once. Given the frequent complaint that TV is already boring, and that new programs are similar to old programs and to each other, isn't it strange behavior to consciously select repetition? Are the jokes on a sitcom still funny when you know the punch lines? Is news still "news" when you re-view it? Is a ballgame still exciting when you already know who won?

RE-VIEWING THE LITERATURE

Apart from a handful of articles in popular periodicals depicting the nostalgia of television reruns, and a smattering of data from the television and film industries about the economics and popularity of re-running old programs/films, little has been written on the phenomena of re-viewing televised content. Zillmann and Bryant (1985) wrote that exploring repeated exposure to the same offerings is "easily the most neglected phenomenon in selective exposure" research (p. 10). They pointed out that although we take repeated exposure to auditory media as a given (we all listen—and choose to listen—to the same songs or pieces over and over again on the radio or on recordings), no one has paid much attention, theoretically or empirically, to the question of repeated exposure to audiovisual materials.

Tannenbaum (1985) suggested that the very structure of the U.S. television industry is conducive to multiple viewings. Multichannel replications of the same program (such as repetitions of popular programs on network and auxiliary channels), same-channel replications (such as PBS stations repeating shows at two or more separate times during the same week), same-day replication (such as multichannel cable systems carrying several of the same shows at different times during the day), delayed replications (television shows repeating themselves during the season or during the summer), and long-term replications (syndicated television reruns) are all common practices, and afford viewers multiple opportunities for multiple viewing of the same content.

Television reruns mean almost pure profit for the independent and network-affiliated stations that air them. Because production costs continue to escalate, most often profits from programs cannot be realized during their initial runs. When a show has "proven its worth," producers can launch it into a second incarnation in syndication. After syndication fees, there is virtually no expense because all production costs have already been paid, and in fact, this is when a typical show will begin to make money. Popular shows bring high advertising revenues: "M*A*S*H," for example, is expected to bring over $250 million in its syndicated lifetime ("Sharing that syndication gravy," 1983, p. 47).

We can also be exposed to similar content more than once because sometimes media content is used, and then recycled, by those within the media industries. Many media critics have discussed this "preoccupation with the media past," and have written about how the same themes, plots and stories re-emerge," sometimes as sequels ('Rocky IV' and 'Jewel of the Nile'), or spinoffs ('The Colbys'), but often as nostalgia" (Stark, 1985, p. A24).

Tannenbaum (1985) pointed out that some of the newer communication technologies have as "their raison d'etre . . . to provide opportunities for endless replications of favorite audio-visual material, including current television fare" (p. 231). The short-lived video disc, and the videocassette recorder were developed "anticipat[ing] a high level of repeated usage for the system to succeed economically. Again, factors such as initial investment costs point in this direction" (p. 232). The technical ability of the VCR to play back programming multiple times posed legal and ethical questions of royalty distribution (Lardner, 1987), and the rampant pirating of material on cassette has led the various media industries in the United States and abroad to reconsider and alter marketing and distribution strategies (Boyd, Straubhaar, & Lent, 1989, p. 41).

It is clear that television offers us the possibility of seeing the same programs over again, and that video enables us to do this even more easily. It is also clear that rerunning material is economical, both for those who distribute it, and for those who consume it. But why do we?

Geertz (1983) discussed a ritual theory of social action. He suggested that much of what we do in life we do like a theatrical performance—ritually and repetitively. Almost any kind of social action is "rooted . . . in the repetitive performance dimensions" (p. 28), and it is in the repetition that we reinforce our connection to our common cultural ties. "Reiterated form, staged and acted by its own audience," concluded Geertz, "makes theory fact" (p. 30).

Televised content has been discussed by some theorists as one manifestation of those common, repetitive cultural ties. Gerbner, Gross, Morgan, and Signorielli (1986) wrote that "television is a centralized system of storytelling. It is part and parcel of our daily lives. Its drama, commercials, news, and other programs bring a relatively coherent world of common images and messages into every home" (p. 18). Anthropologists and folklorists have hypothesized that repetitive storytelling serves the function of reinforcing certain cultural norms and values, and of maintaining the status quo. Gerbner et al.'s empirical work certainly suggests that television serves this function, at least among those who expose themselves most often to TV's messages (Gerbner, 1985; Gerbner et al., 1986). The implication here is that multiple exposures to multiple incarnations of the same messages are functional within a society. Wrote anthropologist Michael Dorris (1988):

> whether these values are initially transmitted by a tribal elder reciting fables around the fire, parents reading their favorite fairy tales to their offspring before bedtime, or by [my children] and me [watching TV] together for half an hour . . . each method creates a shared set of formative impressions, a common cultural shorthand that collapses the barriers of time and experience, a bond that permits old and young to harmoniously inhabit, however temporarily, the same world. (p. 36)

Others posit more psychological explanations for the gratifications derived from exposing one's self to the same material multiple times. Tannenbaum (1985) formulated an emotional arousal model to account for the popularity of repetitive television entertainment: Without much personal investment, TV heightens our emotional excitation and arousal, if only modestly. He suggested that this model can also be applied to "subsequent motivated replications" of the entertainment experience. Viewing visual content more than once may be preferred by people for a variety of reasons that Tannenbaum elucidated, including the comfort of familiarity, the reliability and certainty in seeing what we already know, nostalgia, fantasy, and cultural ritual (pp. 239–240).

Indeed, the literature from the popular press seems to reinforce Tannenbaum's theories. Park (1989) explained that the slew of veteran TV sleuths returning to the airwaves in the late 1980s (along with other genres of popular reruns) shows that "the television networks have a penchant for digging through their closets for the old and comfortable . . . If there's a motto for television, it is: There's comfort in the familiar" (p. 31).

Many recent reviewers have focused on the theory that the TV experience is enjoyable to us because it can fill us with nostalgia, reminding us of easier, simpler times:

> This intangible television experience may be one reason watching television reruns can be such an all-consuming addition. It recaptures within the context of an individual program the memories and feelings you had when you first watched it. Situation comedies are best at re-awakening such memories and feelings because they constituted a weekly experience. Who would have thought that these simple programs, which flashed so briefly then seemingly disappeared, would come back to us now as a sort of time machine? (Saltzman, 1986, p. 23)

Marc (1983) wrote of watching televised content more than once that "the recognizable, formulaic narrative releases the viewer from what became the superficial concerns of suspense and character development" (p. 36). Viewing content with which we are familiar thus brings us feelings of ease and comfort.

Dobrow and Jackaway (1985) found that watching television reruns was a common practice among their sample of people, most of whom also reported that they watched movies more than once, and re-read books multiple times. Respondents reported a variety of gratifications derived from this repeated exposure to familiar content, including nostalgia, sharing content with others, and re-viewing "classics."

But most of those who have written about or studied repeated exposure to entertainment content did so before VCRs became such a prevalent item in the American domestic landscape. Has the phenomenal penetration of VCRs (estimated at more than 60% of all American television households at this writing) changed, exacerbated, or actually increased the extent to which people watch programs multiple times? What might explain why they engage in this behavior, and what might it all mean?

METHODOLOGY

To begin to explore some of these questions, we included several questions about using the VCR for multiple viewings as part of a larger survey. Trained graduate student interviewers conducted in-depth interviews with almost 200 people in demographically diverse parts of the greater Boston area. Interviews were done in a variety of languages (including Spanish, Japanese, Chinese, French, Greek, Creole, and English), and interviewers translated their findings. Respondents were asked a variety of questions about their patterns of VCR use and their demographic backgrounds.

In all, the survey yielded 193 usable interviews. Of the interviewees, 53% were male (n = 103), and 46% were female (n = 90). In general, their demographic characteristics conformed to those already identified by many researchers as typical of the U.S. VCR market (see Levy, 1987): The respondents were mostly young, highly educated professionals or white-collar workers. However, a significant portion of this sample was comprised of people who represent the second wave of VCR owners. Almost one third (28.5%; n = 55) had *not* gone to college, although they were of college age or older. Many were immigrants. (Dobrow, 1989, reported that in many urban contexts, VCRs are often found in the homes of immigrant families.) Sixteen percent (n = 31) stated that their current occupation was some type of blue-collar job. The majority of people in the sample owned at least one VCR (82.9%; n = 160), and most had owned it for 2 years or less (48.7%; n = 94).

RESULTS

What was not surprising about the responses was that virtually each person interviewed reported that he or she used the VCR to view things more than once; what *was* surprising was the variety of programming viewed multiple times, and the similar reasons given for multiple viewings across demographic and ethnic groups.

Movies (rented, purchased or prerecorded videocassettes, or taped off of network or cable television) led the list of programs watched more than once. Of television programs other than movies time-shifted and played back later, surprisingly many respondents reported viewing sports games (when, presumably, they already knew who had won), news programs (which, presumably, were no longer "news"), situation comedies (when, presumably, viewers were already familiar with the jokes), and soap operas (which, certainly, are repetitive enough without multiple viewings).

Many of the respondents also stated that they watched "home movies" (family events such as weddings, parties, bar mitzvahs, etc.) more than once. Several respondents viewed "how to" or educational tapes multiple times. Exercise videos were also watched repeatedly, as were professional training tapes, concert tapes, and music videos. Some people even reported recording favorite advertisements and deliberately watching them time and time again.

Although some of these types of programs—principally educational and training tapes—are clearly designed to be viewed more than once (chefmaster and video star Julia Child told me in a letter that she felt video was the ideal medium for teaching people to cook because they could always rewind the tape and start over; Dobrow, 1987), others are clearly not. What begins to explain this curious behavior?

Education

The reason most often cited by respondents for why they would watch any kind of programming more than once was "I get more out of it the second time around." Many people reported noticing details they had not observed the first time around, or being better able to concentrate on aspects of a show or movie ranging from film techniques to reading subtitles to understanding the symbolism of the plot. "You can pick up on subtleties you missed the first time around, and enjoy it more" said a male respondent. One woman said, "It's like reading the Bible—the more you read it or see it, the more you can find new and unexpected things in it."

People whose first language was not English often stated that they used multiple viewings of American TV shows or movies to help them learn the language. For example, a Japanese interviewee stated, "I like to watch shows more than once. The first time, I concentrate on understanding the English. But the second time and after that, I can relax and enjoy what I'm seeing."

Critiquing one's technique was also a major educational reason cited for watching something more than once on the VCR. A doctor said that he played back videos of himself performing surgery multiple times;

several musicians, dancers, and actors reviewed their performances on video; many athletes watched themselves and their teammates and analyzed strategies. Some utilized the slow motion function of their VCRs to better critique their performance.

Many respondents who were parents reported that their children would watch the same program tirelessly, and parental interpretation (or hope) of this behavior was that the kids were "learning something" from it (see, also, Jordan, chapter 9, this volume). Some people stated quite directly that watching a program more than once was educational because only in the multiple viewings, could they learn things like why a nuance worked, or how a magic trick was performed or special effect or stunt enacted.

Initiation and Solidarity

Initiating a friend, family member, or significant other into one's favorite cultural symbolic environment constituted the second most prevalent response among this sample. Many respondents told of viewing something multiple times to introduce important people to important films, TV episodes, or home movies. "I like to see old classic movies, like *The Maltese Falcon,* with people who haven't seen them," said one man. "Everyone should see it, and that way I'm a part of the experience. That's what you do with cult films."

"I want to share with special people the stuff that I really like. I want them to see my world, and part of it [are] the shows and movies that are special to me, for different reasons. They understand why these things are special, and they understand me," said one young woman.

"It's great to watch these family movies more than once," said another respondent. "You can relive great moments with your family, or great vacations." Another woman stated, "I like to show my family movies, the video of my wedding, to new friends. That way they kind of can participate in it—even if they weren't there." In addition, showing home movies sometimes initiates outsiders into a family: "It's like showing someone my family photo albums, only better. That way, they can really see what we're like," said a female respondent.

Many multiple viewers also said that in repetitive viewing comes a sense of solidarity with others. "I watch Haitian movies a lot—the same ones, over and over," said a male immigrant. "I'm far from home and I can't go back—but I watch these movies and I feel a part of my people." Several other immigrants responded similarly.

Some respondents stated that in watching entertainment programming multiple times, they felt they could develop a relationship with characters. "I feel like I establish a relationship with the actors," said one woman. Another confessed that she watched scenes from soap operas over again because "I just love Travis [a character]: Watching his scenes over again makes me feel like I really know him, and know what he's going to do next."

Others found that watching videotapes over and over with other people gave them a sense of collectivity with one another. "I invite American friends over to watch Chinese films that have English subtitles," said a Taiwanese respondent. "Watching together is a good way to deepen our friendship." An American male reported that he often would re-view taped baseball games with other Red Sox fans. "It doesn't matter if the season's over, or if you already know who won. What's important is watching it with other people who feel the same way you do about the team."

Viewing home movies with family members more than once seemed to serve similar functions for some respondents. Interestingly, many reviewed events with family members immediately after they had happened. "We all sat down and watched the party again," said a woman, "and we felt incredibly close."

Ritualized Viewing and Participatory Viewing

For some respondents, watching programs more than once was a ritual. Some reported viewing the same film or program at set intervals. "Every Halloween a group of us gets together and watch *Psycho*," said one male respondent. "Each year we watch the videotape of our wedding on our anniversary," said another.

Other respondents told of watching the same thing over and over again with children, although not necessarily at certain times. "We have to watch some of the same things, Disney movies, time after time," said a mother. "It's a kind of ritual. The kids are disappointed if I won't watch with them. And we always have to have the same foods and sit in the same places."

Another sort of ritualized viewing was participatory in nature. Many respondents explained that the primary reason they enjoyed viewing things more than one time was that they could memorize lines or songs, and say (or sing) them aloud with the characters. Some like to anticipate punch lines. Many seemed to feel jokes got funnier, or they could better appreciate the humor of a movie or program upon seeing it over. Others said that they felt they knew the characters better. "I like knowing what's going to happen before it happens," summarized a male respondent.

"It's much easier on the brain, because you already know the plot and the lines," said another.

Particularly among younger respondents, several mentioned getting together in groups and re-viewing familiar TV shows or movies. "We all get together, and say the lines with Kirk and Spock [of *Star Trek*]." "They're like stories that you tell over and over again," reported one interviewee of her multiple viewings with friends. "Some stories," reflected one respondent, "are important to remember. Seeing them more than once helps you to remember."

Control Over Content

Many respondents stated that the advantage of watching programs multiple times was actively controlling their viewing environments. "Why should I watch crap when I have what I like on tape?" asked one male. "I prefer to watch the French movies I know than to see other things which I don't know and might not like," said a woman from Haiti.

"Since I already know the show," explained another woman who said she watches many programs repeatedly, "I can zip ahead to the parts I like and skip the boring stuff." Several respondents reported using the remote control to rewind and replay favorite parts of favorite shows or films or instructional tapes multiple times, and to edit out what they did not wish to see.

Making the Ephemeral Last

Another extremely interesting response given by many of those interviewed was that the VCR enabled them to use television as they would use books. "I like to watch things more than once for the same reasons that I like to read books over again," said one woman. "Watching programs more than once is like reading—you know what you like, and you repeat it," another man commented.

Some respondents felt that having programs on videocassette was like building a library. "I have a collection of my favorite shows and movies. When I'm bored or there's nothing on, I can just pop in a favorite," a male respondent stated. This "store it up for a rainy day" philosophy was shared by many of those people interviewed (see Heintz, chapter 8, and Jordan, chapter 9, this volume, for more information on how people are using video libraries).

Many interviewees said that they watched things multiple times because this was an economical way of being entertained. This view was especially prevalent among parents, and among some of the immigrants interviewed, who reported that they watched the same programs or movies over again because there just were not too many choices available to them, or that it was too expensive to import many cassettes from abroad.

Several of those interviewed mentioned that having something on videocassette meant having something tangible, something permanent. "Our wedding has been recorded forever—years from now we can watch it, or our kids can watch it, and it's still there," said a young woman. And a male respondent concluded, "The world is uncertain—you need to know that that movie [that's so important to you] is still there, and you can see it again, and again."

CONCLUSION: FAMILIARITY BREEDS CONTENTMENT

Viewing programming repeatedly serves several different functions. Educationally, repetition can help us to learn. The many people in this sample who stated that they liked to see things repeatedly, or that their children did, are testament to this. VCRs enable those who use them to notice new aspects of something they have already seen, to concentrate on details in a videotape, or to go back and talk about particular aspects with others. The great expansion of the educational video market will no doubt continue, as will the use of the VCR in a plethora of educational settings, from the classroom to the courtroom to the operating room.

But another kind of learning also occurs from watching entertainment programming multiple times. Watching television reruns, or viewing any entertainment programming day after day, year after year, rehearses the same stories and provides us with more than ample opportunity to learn and repeat cultural narratives. Before the advent of electronic media, cultural norms and values were passed down and perpetuated only by the spoken word, and then by the printed one. But visual media provided a unique function—they *showed* us these norms and values, continuously demonstrating forms of patterned behavior. Television experiences, however, were "so transient and personal that viewers yearn[ed] to have them validated. In our society, the true validation of any fleeting audio or visual experience has always been the printed word" (Saltzman, 1986, p. 23). But VCRs may be changing that—we can record television experiences for posterity.

Some media critics have written that the popularity of some television reruns has to do with certain "enduring values" shown with clear, albeit stark, simplicity. "Friendship and family are prized. Violence catches our attention, but only momentarily . . . The only programs that secure a place in the permanent repertory are the ones that confirm the contours of our identity" (Corry, 1986, p. 25). The VCR clearly enables those who so desire to expose themselves over and over again to these same stories, and, as Morgan, Shanahan, and Harris (chapter 6, this volume) suggest, paves the way for cultivation of values within the stories among those who continually expose themselves to them.

Several people in this sample spoke of how VCRs can make programming more permanent. For the repeat viewer, this is not only economical, but in many cases, seemingly symbolic. Repeated viewing is a way of prolonging entertainment, making it last. Said one respondent, "I like to watch things more than once because it reassures me that that little part of the world is still there, just like that—it doesn't change, and it won't ever."

Many re-viewers seemed to like to view old programs because in their dated, unchanging dialogue and styles, they reminded the viewers of "simpler times." Entertainment programs serve as benchmarks in our lives. "The world recalled may never have existed in quite the way it is remembered. Nevertheless, if a situation comedy brings back happy memories, even false happy memories, the effect on the psyche is the same" (Saltzman, 1986, p. 23). Similarly, those people who said that they watched news or personal histories on video over and over again were no doubt remembering what the world was like for a moment captured on tape, and comparing where they are now to where they were then. There is comfort to be had in this, and as Coles (1989) suggested, we often return to favorite stories at different points in our lives and find that the changes in *ourselves* often inspire new interpretations.

"This shadow memory [of the media text] is interactive with individual memory; it provides images that function as personal signifiers . . . and at the same time serves to document and redocument collective experience" (Marc, 1983, p. 135). The familiarity and comfort that we derive from watching the same programs comes not only from knowing ahead of time what is going to happen, but also because the very act of watching them, time and again, becomes an affirmation of collective experience. Much of the data gathered in this study supports this notion: The respondents who reported that they liked to view programs more than once so that they could anticipate lines, plots or songs, and those who said that in re-viewing they could intensify their relationships with individual characters were consciously aligning themselves with a part of their culture, seen and experienced by millions.

Additionally, many respondents said that the primary purpose of watching programs multiple times was to introduce or initiate those who had not seen them. Others stated that video was an important mechanism for sharing personal histories—weddings, parties, family events—with those who were not there. This often seems to happen despite the fact that the experience of watching one's self on video is not always a pleasant one, as Vale (chapter 11, this volume) discusses. Will video eventually replace family photo albums, or make one's personal history more accessible? What will it mean to future generations to have video records of their predecessors?

We often think of television viewing as being a solitary activity, but at least in the case of viewing home movies or reruns of films or television programs among many in this sample, people clearly are using their VCRs as a way to watch *with* other people. There have been conflicting findings on this issue, as Lin (chapter 4, this volume) noted; clearly, more research should follow up on this point. Is there something special about re-viewing content with others, perhaps initiating them into a particular media experience, that encourages group viewing?

Finally, given the ever-increasing range of entertainment opportunities available to people through cable and pay/TV, satellite and video, the fact that so many people actively and deliberately seek to re-expose themselves to familiar content may seem strange behavior at first glance. But let's take another look. For one thing, it challenges the notion that television is watched primarily passively and unselectively. But re-viewing means more than that. The preference for repeated media experiences provides people with a common set of images, symbols and narratives, possibly connecting diverse groups of people through providing them with a basis for common memories. Just how strong and how widespread this behavior is has yet to be tested. But repetitive exposure to repetitive content—both visual and aural—and the act of watching it, again and again, is a ritual that may reinforce our cultural patterns as they've never before been reinforced.

REFERENCES

Boyd, D. A., Straubhaar, J. D., & Lent, J. A. (1989). *Videocassette recorders in the third world.* New York: Longman.

Coles, R. (1989). *The call of stories.* Boston, MA: Houghton Mifflin.

Corry, J. (1986, August 3). Why some old shows never fade. *New York Times,* p. 14.

Dobrow, J., & Jackaway, G. (1985, April). *To boldly go where we have gone before: Toward a paradigm of rerun viewing.* Paper presented at the Popular Culture Association Conference, Louisville, KY.

Dobrow, J. R. (1989). Away from the mainstream? VCRs and ethnic identity. In M. Levy (Ed.), *The VCR age* (pp. 193–208). Newbury Park, CA: Sage.

Dobrow, J. R. (1987). *The social and cultural implications of VCR use: How VCR use concentrates and diversifies viewing.* Unpublished doctoral dissertation, Annenberg School of Communications, University of Pennsylvania, Philadelphia, PA.

Dorris, M. (1988, May 28). Why Mister Ed still talks good horse sense. *TV Guide,* pp. 34–36.

Geertz, C. (1983). *Local knowledge.* New York: Basic Books.

Gerbner, G. (1985). Mass media discourse: message system analysis as a component of cultural indicators. In T. A. van Dijk (Ed.), *Discourse and communication: New approaches to the analysis of mass media discourse and communication.* Berlin: Walter de Guyters.

Gerbner, G., Gross, L., Morgan, M., & Signorielli, N. (1986). Living with television: the dynamics of the cultivation process. In J. Bryant & D. Zillmann (Eds.), *Perspectives on media effects.* Hillsdale, NJ: Lawrence Erlbaum Associates.

Lardner, J. (1987). *Fast forward: Hollywood, the Japanese and the VCR wars.* New York: Norton.

Levy, M. R. (Ed.). (1987). *The VCR age.* Newbury Park, CA: Sage.

Marc, D. (1983). *Demographic vistas.* Philadelphia, PA: University of Pennsylvania Press.

Park, J. (1989, March 5). Veteran sleuths are donning revamped gumshoes. *New York Times,* pp. 31, 38.

Saltzman, J. (1986, July 27). There's a golden glow 'round those grainy reruns. *New York Times,* p. 23.

Sharing that syndication gravy. (1983, August 15.) *Time Magazine,* 47.

Stark, S. (1985, December 29). A nation rushing back to the past. *Boston Globe,* pp. A21, 24.

Tannenbaum, P. H. (1985). "Play it again, Sam": Repeated exposure to television programs. In D. Zillmann & J. Bryant (Eds.), *Selective exposure to communication* (pp. 225–241). Hillsdale, NJ: Lawrence Erlbaum Associates.

Zillmann, D., & Bryant, J. (1985). Selective-exposure phenomena. In D. Zillmann & J. Bryant (Eds.), *Selective exposure to communication* (pp. 1–10). Hillsdale, NJ: Lawrence Erlbaum Associates.

11

Captured on Videotape: Camcorders and the Personalization of Television

Lawrence J. Vale
Massachusetts Institute of Technology

ON VIDEO

The title of Roy Armes' (1988) book, *On Video,* gives it a place in the "On *X*" genre of analytical essays that ranges from Clausewitz's *On War* to Sontag's *On Photography* (two books whose contents share much more in common than their titles might at first suggest). Armes' title, however, has the added presumed advantage of an implied double meaning; to wit, a pun. Not only does this title offer the suggestion that it will be "on" video in the sense of being an extended essay-of-definition *about* video, it also opens the possibility that it will delve into the subject more experientially to ask what it means to *appear* "on" video. This, to the great disappointment of all of us who believe that the title of any profound book should present at least two simultaneous ways to interpret its contents, proves not to be the case. Armes wrote convincingly about the history of video, the social context of video from the point of the professional producer, and the aesthetics of video sound and video image, but did not examine the question of how video is received by those who find themselves captured "on" it. This next task is the one I take up here.

The camcorder, first launched in 1985 as the one-piece descendent of the two-piece video recorder (Marbach, 1985), would at first seem no more than an accessory to the increasingly ubiquitous VCR.[1] Yet, as I

[1] In some cases the camcorder is not an accessory to the VCR but, instead, its substitute, because some users connect their camcorder directly to their television and do not own a separate VCR.

argue here, it is a curious hybrid of many technologies, and is a product that itself seems likely to introduce some new strains. Certainly, in its domestic applications,[2] the "use value" of a camcorder is close in many ways to that of the home movie camera, the technological dinosaur that it is fast replacing (Chalfen, 1987; Hughey, 1982; Kealy, 1981). Still, as an electromagnetic alternative to the photochemical techniques used in film processing, the camcorder and its videocassettes are technologically quite distinct from the principles that undergird the earlier medium. In this sense, the camcorder is evolutionarily more closely related to the audiocassette recorder and certain other forms of sound recording than it is to film (Armes, p. 9). There are other important linkages, technologically distinct yet socially and culturally interwoven. As a social tool, the camcorder is certainly close to the still camera and the world of home snapshot photography, just as the videocassette—when used as a mode for the interpretation and home storage of recorded family history—bears some relation to the home photo album, as well as to earlier traditions of the family Bible and the painted ancestral portrait (Beloff, 1985; Chalfen, 1983, 1987; Hirsch, 1981; Jacobs, 1981; Lesy, 1980). As a means to produce near-instantaneous reproduction of images, the camcorder is cousin to the Polaroid camera, and the camcorder/VCR/TV ménage-à-trois is the thriving successor to Polaroid's "Polavision," their abortive attempt to market a system of "instant home movies" displayed on a nontelevision monitor (Hughey, 1982; Olshaker, 1978). This last media misconnection suggests also that to begin to assemble the meanings that are associated with the camcorder and the home videos that its users produce, one must consider the importance of the novel intersection that has been created between what Chalfen (1987) called "home mode communication" and the external mode programming that is more usually associated with television sets. How important is it—and in which ways—that home videos are viewed on *television*? Thus, although the camcorder is closely related to many other media, it is not an exact substitute for anything, even home movie equipment. As a new media technology, it shares many things with the introduction of other new technologies but, nevertheless, it differs—technically, aesthetically, socially, and culturally—from all that has preceded it.

All of the recent commentators on the various modes available for using media to record family history have stressed the ways that this

[2] Video cameras and camcorders are, of course, used in many settings and for many purposes. They are used for diagnostic and documentary purposes in many professions, and are becoming an increasingly important component of television production. In addition, video, as a medium in itself, is an increasingly popular art form. My central concern here, however, is with the social uses and cultural meanings of video in the home context.

history is edited—both consciously and unconsciously—to reflect positively on lives being lived. A column of helpful hints in *Popular Photography,* entitled "Family Photography as a Sacrament" (Hattersley, 1971) makes clear that this selective editing should be an explicit ingredient in good photography:

> Use good judgment in picking the time and place for pictures. Children don't worry much about the when and where, and are usually raring to be photographed. With adults, however, it's a different thing, for their worries and concerns leave them feeling pretty bedraggled much of the time. It isn't good to photograph Dad right after a hard day's work, Mom when she's been hassled by a bill collector, and Junior after his girl has given him the brushoff.
>
> Be on the lookout for those golden hours when everything seems to be going right for everyone in the family. It may be the right time on a lazy Saturday morning after a late breakfast, or for the half hour just after the finish of the family's favorite TV show. After finding these sweet times, make sure your approach to photography makes them seem even sweeter. (p. 108)

This set of saccharine suggestions would seem to fit well with what most scholars of "home mode" photography have observed. According to Jacobs (1981), photo albums "are constructs that propose positive histories. . . . With snapshots we become our own historians, and through them we proclaim and affirm our existence" (p. 104). Chalfen (1987) noted wryly that:

> Future anthropologists, if they studied our culture from home photo albums alone, would probably conclude that this breed of man lived mostly as Christmas, indulged in a ritual with colored eggs at Easter, graduated from institutions frequently, celebrated birthdays mostly while young and had lots of small animals. Further, they would conclude, children were usually fresh-scrubbed, and spent a great deal of time standing around squinting into the sun. (pp. 170–171)

Halla Beloff (1985) commented that "We are obliged to show what happiness we have experienced, what friends deserved, what ambiences and positions achieved" (p. 190) and concluded that "What is shown is the past that we can be sentimental about" (p. 196).

"What kinds of information" Chalfen (1987) asked, "are being transferred from generation to generation between the covers of a family album, in cans of home movies, or in videotape cassettes?" (p. 2). Chalfen's answers, and those conclusions proffered by other studies of various media, confirm a certain consistency across these three media both in terms of subject matter and in terms of attitudes toward it: there are more images of young children than older ones; more of first born than of later births, more emphasis on achievements than defeats, more depictions of vacations than vocations. Armes (1988) continues in his criticism of the superficiality and artificiality of home video:

> [V]ideo is a technology symptomatic of the public role given to images in a capitalist society; it records aspects of the surface of life, but it embellishes, prettifies, as it records. . . . [The video camera] is openly, transparently, both an instrument for celebrating what *is,* rather than what could be achieved by social change, and, at the same time, a machine for making life seem more pleasurable than it is. (p. 197)

Although Armes is in clear agreement with those who suggest that home video (or, more precisely, those persons who make them), like earlier modes of domestic history gathering, tends to put a positive gloss on the events of life, it does not necessarily follow that the camcorder really does act as "a machine for making life seem more pleasurable than it is." Even these edited records of life may yield emotions that are far from pleasurable, due both to the process of videomaking and the resultant product.

To test this hypothesis about attitudes toward the process of videomaking with home camcorders, this chapter employs data from a set of nearly 200 interviews with VCR owners conducted in early 1989 by graduate students from Boston University's College of Communication. In the course of each interview, each person who reported having watched him or herself on video was asked to comment on the experience. In addition, those in this sample who said that they owned, rented, or had access to a camcorder or videocamera were asked not only to identify the kinds of things they chose to videotape, but also to state whether there were certain things that they would *never* want to portray on videotape. The responses to these questions suggest that, even if the finished video products are sanitized selected views of atypical situations, these products may have little to do with enhancing life's pleasures.

VIDEO DO'S AND VIDEO TABOOS

More than two thirds of this sample of VCR owners reported having watched themselves on video, a reminder that the social impact of camcorders goes far beyond the statistics about camcorder ownership penetration. Although camcorder ownership in the United States has grown rapidly and at this writing stands at about 7 million—equal to about

7% or 8% of all television households (*Videomaker;* Electronics Industry Association, December 1988, cited in Collins, 1989)—the prevalence of VCRs—approximately seven times the penetration rate of camcorders—has already greatly extended the accessibility of the video products created by home camcorder technology.

Before turning to a discussion of the questions about how respondents said they reacted to viewing themselves on television videos and what camcorder users would never want to videotape, some comment seems useful about how camcorder use may be related to other patterns of media use. Of 49 respondents who owned, rented, or had access to a camcorder or video camera, nearly three quarters said they had at some point owned a still camera and made photo albums.[3] Of the remainder who did not make photo albums, nearly half said that they had made photo albums before they began using a camcorder. Although the sample size for these interviews is too small to draw any firm conclusions, it does seem likely that the home-produced videotapes may, in many cases, supersede the family photo album as the primary repository for domestic documentary history.

About 40% of those using camcorders reported having owned a Polaroid camera at some time, a proportion consistent with the range of instant camera penetration figures over the last decade (Wolfman, 1980; Photo Marketing Association, 1985, cited in Wolfman, 1987). Approximately the same proportion stated that they had, at some time, owned an 8 or 16mm movie camera. Not surprisingly, the correlation between camcorder ownership and previous movie camera ownership is high. The percentage of camcorder users reporting previous movie camera ownership is about one third higher than the maximum penetration rate of movie cameras in the United States (achieved in 1977). Indicative of its increasing obsolescence of home moviemaking equipment—80% of which still relied on soundless systems as late as 1980 (Wolfman, 1981), it is estimated that only 3% of American households were still using movie cameras in 1988, down from 12% in 1980 (Wolfman, 1980, 1988). In other words, current camcorder use is estimated to be nearly three times greater than current movie camera use.

Of those who said they had watched themselves on video, most reported that the videos showed them at social events with family and friends—primarily weddings, parties, and birthdays. The next most commonly mentioned subject matter focused on the workplace. Other subject matter frequently mentioned included performances such as recitals and skits (10%), school-related activities (8%), and sporting events (8%). What

[3] This figure for still camera use coincides exactly with the national average—73% of households (Wolfman, 1988).

is perhaps surprising is that only 5% of those who said they had watched themselves on video reported that this video was taken during a vacation. This surprisingly low proportion may be explained, in part, by the fact that only 37% of those who said they had watched themselves on video personally owned, rented, or had access to a camcorder. Of those who themselves *owned* a camcorder, 42% reported using it on vacations.

When asked to identify anything that they would *never* want to film with a camcorder, only 17% of the camcorder users said that there was nothing that they would not want to film. An additional 17% gave noncommittal or uncertain responses. By way of contrast, a full two thirds of the sample, without any prompting from the interviewers, had a clear idea about situations that they would not want to film (cf. Chalfen, 1984). Not surprisingly, perhaps, the responses of camcorder users centered on the Freudian duo of sex and death. Repeatedly, both male and female respondents stated their refusal to let the camcorder into the bedroom. In addition to prohibitions on filming sexual relations and home-produced pornography, several respondents (mostly East Asians) said that they would never wish to film someone sleeping or just awakened. A few respondents thought to extend the camcorder ban to the bathroom as well. A substantial number of respondents stated that they would not want to film situations connected with death: no "funerals"; "no wakes, no wills"; "nothing bad—like the report of a death." In at least one case, the ban on funeral filming was based on actual intergenerational experience, directly prompted by a woman's distress at her mother's cinematic treatment of her own mother's funeral. Another respondent, an Hispanic obstetrician, said that although he regularly helps film childbirth, he would never want to film the moment when "the baby learns [about] death."

Whatever its potential to document individual and family experience, most camcorder users wished to impose clear spatial and ethical boundaries. Their responses suggest a powerful underlying fear of both the intrusiveness of this technology on closely guarded realms of privacy and a further fear of the camcorder's potential to perpetuate or revive unpleasant memories. Even more frequently than mentions of sex or death, respondents stressed the more general concern that the camcorder not be used to expose undesirable parts of private life. Many camcorder users insisted that they did not want to film what they variously called "intimate life," "very private life," "anything private or personal," "private things," "anything private or concerning the private life of someone," "personal things," and "anything personal." Some respondents, perhaps out of this same kind of discomfort, preferred to give a humorous response. One man said he would never film his "stepmother in a swimsuit"; another proposed a special ban on depicting his mother's cooking. Humor, as always, remains intimately tied to anxieties. Other camcorder users dispensed with qualifying words like "swimsuit" or "cooking" and made clear their desire to exclude completely certain people from appearing in their videos. One woman singled out her stepfather for exclusion, whereas another, more categorically, said she would not wish to film "people who I don't like." A college student preferred to edit out any activity associated with her relatives, "like, family things, you know."

In contrast to this pervasive and conscious desire to edit out unpleasantries and limit visual access to personal and private affairs, some members of one category of respondents seemed more concerned with using their camcorders to document something closer to a more balanced interpretation of their positive and negative experiences. This category of camcorder users is comprised of recent immigrants to the United States, who use their videos as a means to communicate their American experiences to those still living in the "old country" (an ancestral home presumably well-endowed with the proper VCR receptor). Although the number of individuals in this sample who are engaged in First World/Third World home video transfer is too small to derive anything beyond anecdotal evidence, alternative patterns of camcorder use and significance seem quite possible (see Dobrow, 1989). One Haitian man, for example, reported that his primary use for a camcorder was less to film "family or social events" than to depict "political events or to record some aspect of American society such as the homeless." He wished to do this, he said, "because some fellow countrymen [in Haiti] still believe everything is in gold in the U.S." A Cambodian refugee, less concerned perhaps with documenting the downside of life in America, also differs from the mainstream of home video tendencies when he says that he would never wish to film "things that are not important to my family." Although it is not clear exactly what would constitute an "important thing" to him, he too seeks self-consciously to document his family's accommodation and adaption to a new culture. For him, the camcorder is used "to record our life in the U.S. to send to our family abroad to see how we progress." Whether these documents of progress are accompanied by snippets of setbacks is uncertain. What seems clear is that the exported home videos are intended to be used as evidence. Each of these camcorder users describes ways that they edit their lives for television, but video lives may be edited to serve many different purposes.

In accord with previous studies that have examined the content of home movies and home snapshots, this investigation of home video suggests that image content is selectively cultivated to favor special events and gatherings not part of the normal routines of daily life. There are clear—if often unspoken—boundaries of acceptability, and there is a general agreement among camcorder users that the machine must not be allowed to intrude too closely on private or personal life of those who are being filmed. How, however, are these restraints received by those who see and hear themselves depicted on video? Does the camcorder user's proclaimed desire to keep the camera away from the most private realms of experience permit those "captured" on video to be seen only as they would like to be seen?

THE TELEVIDEO SELF

To get at this issue, 122 VCR users who reported having seen themselves on video were asked "What was it like to see yourself on TV?" Interviewers were asked to try to record each respondent's exact words. These responses were then coded in several ways. First, key words from the responses were coded and divided according to type of reaction denoted: positive, negative, and neutral/mixed feelings. Next, for further analysis, responses were divided into those persons who had seen themselves on video and who themselves used a camcorder and those who reported having seen themselves on video but did not personally use a camcorder. Finally, the responses were subdivided according to sex.

The first striking observation was that remarkably few people—only 15%—expressed opinions that could be categorized as "neutral" or "mixed feelings": This was clearly a subject where people reported clear views, whether positive or negative. Moreover, of this 15% who could not be easily categorized at one extreme, 39% responded by expressing marked ambivalence, characterized by strong but conflicting feelings. Thus, out of the whole sample, only 9% described the experience as being "no big deal" or were unable to articulate any particular response.

Of the remaining 91% who reported more clearly resolved sentiments, nearly two thirds expressed a negative reaction to seeing and hearing themselves videotaped on television. The negative reactions ranged from the rather mild—35% used the words "weird," "strange," "funny," "odd," or "peculiar" to describe their experience—to the thoroughly traumatized—22% used words such as "horrible," "awful," "scary," "depressing," and "grotesque." The rest of the negative responses could be grouped somewhere in the more middle regions of tolerated unpleasantries. Many respondents used the words "uncomfortable" or "embarrassed" to describe their reaction in more general terms, whereas others zeroed in on more specific aspects of their own behavior that served as the source of the discomfort and embarrassment. Some described their distress at finding their body movements appeared "awkward," "goofy," or "foolish," whereas others bemoaned certain more static aspects of telegenic inadequacy—they saw themselves as "ugly," "fat," or "old." One man said, "It was shocking—I realized I'm really getting bald." Another noted that "It shows every wrinkle and fat." A third concluded that "ever since the first day I saw myself on video I've always considered that I have a face for radio." A fourth respondent said: "I have never been satisfied with what I see. Although I knew I was an ugly guy, on TV I am 10 times uglier." Another summed up: "Sad and depressing."

Even more than men, women who were asked to comment on the experience of watching themselves on television reported seeing evidence of their own physical inadequacy. Statements included such things as "I looked fatter than I had expected," "I didn't like myself in a bathing suit," "I looked 10 pounds heavier," "It was sad and disappointing—I was too pale," "Embarrassing, and I thought I looked older than my age," and "Horrible—I didn't look good. You look fatter on TV." These comments about visual self-perception, however, constitute only one part of the way that discussion about the experience of video was structured.

Although many respondents answered in a general way that did not identify which component of their televideo selves was the greatest source of pleasure or displeasure, responses ranged across three aspects of the video experience—the perceptions of audio, visuals, and kinesics. A man, one of many who reported the experience as "embarrassing," identified the source of discomfort in terms of audio rather than visual experience: "I never thought that my voice was that high-pitched." A woman remarked, "It was funny to see me and *hear* me. Finally I realized that I had a false image of myself." Other respondents were most disturbed by kinesic aspects: A woman related her embarrassment at realizing she "put [her] pelvis up front too much." Another, describing the experience as "weird," said that "It's funny to see yourself on TV because you do not think you act that way." And one man reported experiencing the visual, audio, and kinesic aspects of video in terms of a triple threat of unpleasantries. "It was really strange. It wasn't so much seeing the video, but my mannerisms and the way I spoke. And I wasn't thrilled about the shirt I was wearing, although it was one of my favorites."

As to the minority of respondents who felt more positively about the experience of seeing themselves appear in a video, there were three basic tiers of response. The least enthusiastic yet still positive group of respondents (24% of the total number of positive respondents) were coded according to their consistent use of two words: "interesting" and "educational." A second, somewhat higher tier of positive response, also could be coded according to the repeated use of only two key words: "amusing" and "fun." This group, clearly entertained but reporting no great thrills, constituted the largest proportion of the positively inclined part of the sample—38%. Only 35% of the positively inclined respondents used words such as "great" or "exciting," although a few were totally enraptured. One man commented: "I find it a real thrill to be able to see myself on say a TV set which is so alien from us; we are attached to TV but at the same time apart. It's real—you become a part of TV life," whereas a woman said: "I suppose orgasmic isn't something you want to write on your survey—it was . . . very special." Such comments were clearly exceptional, however. Out of this whole sample of VCR users who reported experiencing themselves on video, only 11% of them regarded the experience as overwhelmingly positive.

The perception of the televideo self as an unpleasant experience is even more marked if one looks only at the portion of the sample who reported seeing themselves on television but who did not themselves ever use a camcorder. Although it may not be surprising that those who are presumably in closer contact with camcorders (and who—in many cases—invested $1,000 or more in purchasing one) would be more comfortable with this technology and feel more positively about its use, it is nonetheless striking to observe the extent of displeasure that this experience seems to cause most noncamcorder users. Although only 30% of the total sample reported viewing themselves on TV as any sort of positive experience, this approval rating drops even lower if one simply looks at the responses of those who do not own, rent or have access to a camcorder. Among this group, the approval rating is only 25%. Among women, the reported approval rating was even more miniscule—less than 7%. Taken overall, only 5% of the noncamcorder-using subset reported the experience in overwhelmingly positive terms, not one of them a woman.

CONCLUSIONS

Although most accounts of home media use have stressed the ways that it is used to produce highly selective positive interpretations of life, this data on camcorder use and misuse and the reactions to viewing home video suggests a considerably less rosy picture. Even though most camcorder users claim to proceed only with extreme caution as they prepare to videotape the private lives of their subjects, this does not seem to have prevented a largely negative response to the experience of being videotaped. It would seem that the perception of excessive intrusion into personal realms is not limited to the boundaries of the bedroom or the bathroom, but may be found even in some of the most casual exchanges. Even where videotaping is tolerated or actively advocated, a certain ambivalence may remain. In a 1985 interview, one man described the act of videotaping his family reunion as "a cleverly disguised method of crowd control" (Waters, 1985, p. 53). In the 1989 survey of VCR users, one woman, herself a camcorder user, said that the experience of seeing herself on video was "really awful" because "most people see you from different angles" which made her "very self-conscious." Moreover, she added, her "whole family feels that way." Nonetheless, she observed, "they still want to see themselves." Another woman—who had not used a camcorder herself—also described the experience of seeing herself on video as "awful" but noted that, since "everybody else" looked just as terrible, she "didn't feel so bad." At present, any overarching social or psychological assessments of the reasons for this ambivalence, this tolerated displeasure, must remain largely speculative. Yet, certain perceptual aspects of the televideo self seem logical outgrowths of camcorder technology and the social ritual of video replay on television.

Surely there is some appeal to the interface between home video and the television set, because the act of watching home video on television extends the personal and group flexibility made possible by the VCR.[4] Whereas the VCR enables individuals to shift broadcast timetables to fit their own schedules and allows them to choose from an increasingly diverse range of non-network content, the camcorder/TV interface gives them the ability to create their own programming in terms of content as well as genre. In so doing, the phenomenon of "time-shifting" is applied to personal and family history: Like the VCR itself, the camcorder helps advance the illusion that past events can be brought forward and rendered convincingly in the present.

Although time-shifting was, perhaps, always inherent in the act of looking at family photo albums, this flexibility and personalization is now

---

[4] To date, the verb "to film" seems to encompass the act of making a video as well. Although the verb "to videotape" seems also to be used frequently, it is treated—by the nonprofessional at least—as virtually synonymous with the verb "to film." Conversely, there seems a tendency to refer to the finished product as only as "a video" rather than as a "film" or even as a "videotape." This idea of "videotape" as a noun seems to refer primarily to the blank videocassette before the camcorder user has acted on it. This points to the curious tripartite existence of the videocassette (shared with its predecessor the audiocassette): it is a (a) format for consuming prerecorded and packaged data that is borrowed, rented or purchased (b) format for personalizing this prerecorded and packaged data not only by selecting it, but by time-shifting it, editing it, and storing it (c) format for home-producing personalized equivalents, approximations or alternatives to pre-recorded fare. It will be interesting to see how long it will be before "to video" and/or "to videotape" gains a separate currency as a verb. Perhaps, though the underlying technologies of film and video are very different, there is such a great deal that is similar about the physical and social acts of holding and using the camera, that the new word may be slower to catch on than might otherwise be expected.

paired with the authority of the medium of television. Although home movie viewing required the assembly of often awkward systems of projectors, sound equipment and free-standing screens, with home video the new technology becomes experientially consolidated into the TV/VCR. No longer projected onto a screen on a shaft of light—with sound behind and picture out front—the televideo family image seems to emerge outwards from within the television box, emanating from the same mysterious source as the nightly news or the latest soap opera episode. As Armes observed (1988), "When tapes are viewed, the hierarchical distinction between a multi-million dollar feature film and a domestic recording of a family wedding is erased" (p. 83). A single screen is now shared by multiple media. The seeming self-sufficiency of the video cassette, hidden deep within the interstices of the TV/VCR media complex, would seem to yield at least some degree of conflation—or at least confusion—in a culture that does not now tend to differentiate between "watching TV" and "watching the VCR." In those cultures where the village VCR actually preceded the reach of broadcast television the effects of hierarchical uncertainties might be even greater.

The infiltration of home-produced video into the television world of network-produced broadcasts and studio-produced films seems likely to affect home video's meaning. On the one hand, there is a significant sense of empowerment gained through association with television; on the other hand, the family subjects are implicitly placed up for comparison with professional actors who are even more self-consciously coiffed, made-up, and scripted. Does this, perhaps, account for some part of the widely reported distress at seeing our televideo selves?

Most forms of broadcast television and professional film involve significant amounts of rehearsal and editing before the material is released for distribution. Home video, by contrast, tends to be "live" and largely undigested. Unlike still photographs, where it is the usual practice to throw out bad snapshots and reprint only the most flattering ones for wider distribution, with home video this selection process is more difficult, both technically and conceptually. Given the temptation to review the results of a videotaping session immediately—80% of respondents reported viewing the work "immediately" or "as soon as possible"—and the further tendency of the vast majority to show the video in a group setting, there is little opportunity for editing, even in the unusual cases where the person in charge of the videotape's contents might actually be interested in taking the time to do this. Although the most significant forms of "editing" happen in the process of deciding what to videotape in the first place, after-the-fact editing is not a simple task. Although it may not be technically difficult to erase undesirable parts of the tape, the nature of the moving image makes diagnosing and ameliorating undesirability an extremely complex activity. In the multifaceted context of captured kinesics, it is not easy to edit-out all ill-considered remarks and uncharacteristic grimaces. Video gets uncomfortably beyond the pose and undiplomatically preserves the casual comment, made, perhaps, by someone still clinging to the illusion that cameras photograph but do not "listen." This last aspect—the lack of an audio self-consciousness—may indicate, in part, a sort of "media lag," wishful or otherwise. Nevertheless, even more than in the case of still photography, video may reveal "too much."

At the same time that viewing a videotape may produce discomfort because of its unwanted intrusion into private realms, there is also the related danger that the video rendition of life may be accepted as an alternative to memory, rather than as its supplement. It would seem plausible that the meaning of events is altered by the immediateness of the experience of video replay; reflection occurs in response to the video, rather than to the totality of the event. Participants captured on video are asked to accept the sights and sounds of an event shot from a point of view that may differ significantly from that of their own perceptual experience, an alternative perspective that is then given the added legitimacy of portrayal on television. Also subordinated, or at least disrupted, is the cultivation into memory of other sensory experiences—the tastes, smells, and touches that are even less directly recordable by video technology. Still, at least with these three kinds of sensory data that the videotape does not attempt to replicate or represent, there remains a greater latitude for letting each individual's memory intervene; the visual cues to remembered tastes, smells, and touches may be quite valuable.

Nonetheless, taken overall, the seeming "real"-ness of the videotape medium can encourage a dangerous substitution. "Is it live or is it Memorex?" becomes "Is it life or is it Memovid?" and the consequences of being unable to distinguish become ever greater. In some cases, the temptation toward instantaneous video playback may cause video to not only serve as an important way to memorialize an event but may actually become the event itself: the second half of a party can be the video view of the first half. As one man put it, "If we record a kid's birthday party everyone runs into the other room to see themselves and it breaks up the party. I don't like it because the taping becomes more a part of the event than the event itself" (Stocker, 1988, pp. 54–55).[5] Video's ability to

---

[5] Video "instant replay" has also had a controversial impact on professional sports, where play is interrupted until such time as what has just happened can be reviewed and assessed. The controversy occurs not simply because the audience has a chance to "relive" the moment that has just passed but because game officials rely on this technology to overrule or reinterpret these events, which thereby transforms them. The assumption that the videotaped instant replay should be accepted as a more "accurate" basis for judgment than the eye of the appointed human authority seems significant. Where there are clear "rules" that the videocamera is asked to help adjudicate, there is some justification; as an example of a tendency to abandon the authority of one's own experiential vantage point, this seems part of a disturbing trend.

transform "the event itself" thus may occur not only because of the intrusions of the videomaking process, but also because of the intrusion of the product. At all times, social situations are altered and cultural meanings transmuted. As Armes (1988) commented, "Electro-magnetic recording in particular can take our most intimate situations and, while apparently preserving them, simply turn them into mere information flow, the very opposite of lived experience" (p. 5).

Many of the most interesting questions about the meanings of camcorders and home video will require close observation and analysis over a long period of time. What will it mean for a new generation to hear and see the movements of unmet ancestors? How are camcorders being used differently in different cultures? As photography continues to lose some of its authority in the face of digital retouching processes (Lasica, 1989), will we continue to regard video as a superior form of cultural authenticity? Benjamin (1936/1969), in his classic discussion of the loss of aura in art in an age of mechanical reproduction regards this loss, in part, as a positive shift, a welcome and democratizing liberation from tyranny of upper class high culture. What about this new form of reproduction, even more mechanical than film? Does the camcorder, currently far more expensive to own than any other media with which it is compared, represent a return to mediated elitism, a privileged but illusory personalization of television? Is it possible that the camcorder brings us only the worst of all worlds: a new form of reproduction that is both aura-less and anti-democratic?

REFERENCES

Armes, R. (1988). *On video.* London: Routledge.

Beloff, H. (1985). *Camera culture.* Oxford: Basil Blackwell.

Benjamin, W. (1969). The work of art in the age of mechanical reproduction. In H. Arendt (Ed.) & H. Zohn (Trans.), *Illuminations* (pp. 217–252). New York: Schocken. (Originally published 1936)

Chalfen, R. (1983). Exploiting the vernacular: Studies in snapshot photography. *Studies in Visual Communication, 9*(3), 70–84.

Chalfen, R. (1984). The sociovidistic wisdom of Abby and Ann: Toward an etiquette of home mode photography. *Journal of American Culture,* 7(1–2), 22–31.

Chalfen, R. (1987). *Snapshot versions of life.* Bowling Green, OH: Bowling Green State University Press.

Collins, M. (1989, January 19). Latest newshounds: Video camera buffs. *USA Today,* p. 3D.

Dobrow, J. R. (1989). Away from the mainstream?: VCRs and ethnic identity. In M. Levy (Ed.), *The VCR age* (pp. 193–208). Newbury Park, CA: Sage.
Hattersley, R. (1971, June). Family photography as a sacrament. *Popular Photography,* pp. 106–108.
Hirsch, J. (1981). *Family photographs: Content, meaning and effect.* New York: Oxford University Press.
Hughey, A. (1982, March 17). Sales of home-movie equipment falling as firms abandon market. *Wall Street Journal.*
Jacobs, D. (1981). Domestic snapshots: Toward a grammar of motives. *Journal of American Culture, 4* (1), 101.
Kealy, J. (1981, July 12). Will videotap [sic] systems replace home movies? *New York Times.*
Lasica, J. D. (1989, January 2). Pictures *don't* always tell truth. *Boston Globe,* pp. 29–30.
Lesy, M. (1980). *Time frames: The meaning of family photographs.* New York: Pantheon.
Marbach, W. (1985, December 30). Video's new focus: A small camera makes a big difference. *Newsweek,* pp. 56–57.
Olshaker, M. (1978). *The instant image.* New York: Stein & Day.
Sontag, S. (1977). *On photography.* New York: Farrar, Straus & Giroux.
Stewart, D. (1979). Photo therapy: Theory and practice. *Art Psychotherapy, 6*(1), 42.
Stocker, C. (1988, December 8). Camcorders zooming in on family life. *Boston Globe,* pp. 49, 54–55.
Waters, H. (1985, December 30). The age of video. *Newsweek,* pp. 44–53.
Wolfman, A. (Ed.). (1966, December). *Photo dealer: 1966 annual statistical report: The photographic industry in the United States.*
Wolfman, A. (Ed.). (1980). *1979–1980 Wolfman report on the photographic industry in the United States.* New York: Modern Photography.
Wolfman, A. (Ed.). (1981). *1980–1981 Wolfman report on the photographic industry in the United States.* New York: Modern Photography.
Wolfman, L. (Ed.). (1988). *1987–1988 Wolfman report on the photographic and imaging industry in the United States.* New York: Diamandis Communications.

Author Index

Page numbers in *italics* show where complete bibliographic references are given

Subject Index